The Spirit of Naval Aviation

The Spirit of Naval Aviation

Foreword by
CAPTAIN WALTER M. SCHIRRA, JR., USN (Ret.)

Text and Historical Research
M. HILL GOODSPEED

Featuring
Photographs by
CHAD SLATTERY

Naval Institute Press
Annapolis, Maryland

First Edition
Editor-in-Chief: Captain E. Earle Rogers II, USN (Ret.)
Photography: Chad Slattery
Art Direction and Design: Nancy S. Lichtman
Text and Historical Research: M. Hill Goodspeed
Contributing Editors: Harold Andrews; Barrett Tillman;
Robert L. Lawson; Captain Stephen T. Millikin, USN (Ret.);
Captain Rosario M. Rausa, USN (Ret.)
Digital Design: Jessica A. Barnard, Leslie M. Geiger, Don C. Francis
Jacket Design: Leslie M. Geiger, Nancy S. Lichtman
Jacket Photo: Chad Slattery

Library of Congress Cataloging-in-Publication Data
Goodspeed, M. Hill
The Spirit of Naval Aviation / M. Hill Goodspeed.
p. cm.
U. S. Naval Institute Edition
Photographs by Chad Slattery
1. National Museum of Naval Aviation
I. Rogers II, E. Earle II. Title
1997
Library of Congress Catalog Card Number: 96-70028
ISBN: 1-55750-8364

Printed by Gannett Offset-Pensacola in the U.S.A.

Acknowledgments

Dedicated to Mr. Richard L. Joutras.

The idea to publish this book came from the creative mind of Naval Aviation Museum Foundation member Mr. Richard L. Joutras. Not only did Mr. Joutras conceive of *The Spirit of Naval Aviation*, he also generously provided the seed money to launch the project underway. We are profoundly grateful to him.

Grateful recognition is also extended to the following individuals for their labor of love in ensuring that this volume was accurate, arranged in a logical and an appealing manner, and reflected the grandeur and beauty of the National Museum of Naval Aviation, its aircraft, artifacts and displays: Harold Andrews, Technical Advisor/Contributing Editor, *Naval Aviation News*; Robert L. Lawson, Editor, *The Hook* (1977–1991); Barrett Tillman, Naval Aviation historian and author; Captain Rosario M. Rausa, USNR (Ret.), Editor, *Wings of Gold*; Captain Stephen T. Millikin, USN (Ret.), Editor, *The Hook*.

Following page: *Encapsulated within this small world of electronic switches and dials, Navy astronauts Charles "Pete" Conrad, Joseph Kerwin and Paul Weitz blasted into orbit enroute to their rendezvous with Skylab on May 25, 1973. The three naval aviators were destined to spend 28 days aboard the space station.*

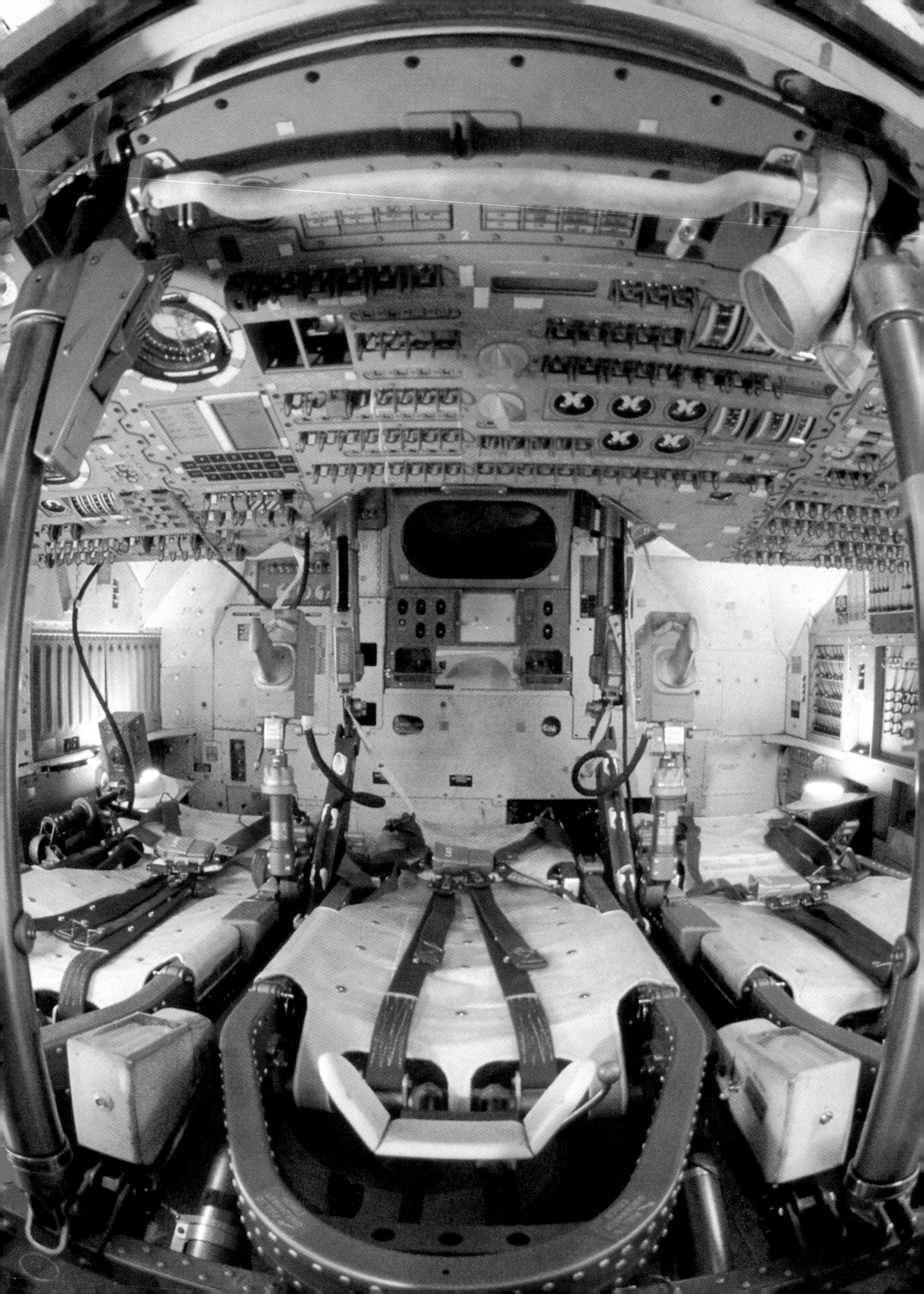

Contents

Foreword

The magnetism of the wonderful spirit of Naval Aviation with its rich heritage and proud history compelled me to earn my cherished wings of gold. It prepared me for and propelled me into astronaut training and guided me safely and successfully through my Mercury, Gemini and Apollo space missions.

That same spirit, so important to me and others in Naval Aviation, has been captured in a most spectacular and dramatic fashion in this exciting book, *The Spirit of Naval Aviation*. A combination of dramatic color images of Museum aircraft from the National Museum of Naval Aviation and historical photographs, *The Spirit of Naval Aviation* captures the magical flying machines in which Naval Aviation has soared for more than 80 years.

From the behemoth NC-4 flying boat, the first aircraft to fly across the Atlantic Ocean, to the warbirds of the Pacific, to the sophisticated jets of the modern era, *The Spirit of Naval Aviation* traces the exciting story of wings of gold. For those who have visited the National Museum of Naval Aviation, this book will bring back pleasant memories. For those who haven't experienced the majestic setting, get ready for a fascinating journey through which you too will feel the spirit of Naval Aviation as seen at the National Museum of Naval Aviation in Pensacola, Florida.

Walter M. Schirra, Jr.

Visitors are greeted by an F-14 Tomcat, which even in silence, is a symbol of raw power.

Thin strands of sunlight mark the arrival of a new day at the National Museum of Naval Aviation. As the sun begins its slow trek across the eastern sky, its rays illuminate an F-14 Tomcat, poised atop its perch as if prepared to roar into the cool morning sky.

Enveloping the Blue Angel Atrium, a halo of sunlight outlines the silhouettes of four A-4 Skyhawks and reveals a cast of "Yellow Perils" sitting patiently as if awaiting the arrival of fledgling aviators.

As the morning's fingers of light reach ever further more treasures are revealed. Some of the aircraft are curiously fragile, seemingly unable to cope with Mother Nature's unforgiving skies, while others project such power that they seem beyond man's control. They sit in silence now, the roar of their engines remembered only by times long since past.

Soon life will come. There are the young, who swarm around aircraft in wide-eyed amazement, and the old, who distantly hear the call "Pilots, man your planes" or look with fond remembrances at a maze of pistons and cylinders upon which their hands once worked.

These people are a blend of past and future, symbolic of paths once taken and flights not yet made. As curious hands wander across fabric-covered wings and wooden struts, and eager eyes peer into the depths of cavernous jet engines, there is a tremendous sense of patriotism, reverence and wonder.

Following page: *The N2T trainer stands in silent tribute to the triumph, disappointment, fear and exhilaration experienced by student naval aviators.*

Though separated by time, aviators are united through understanding the enduring and familiar expression of victory.

A rainbow of colors dances across the fuselage of one of Naval Aviation's early jets, the FJ-1 Fury.

The faces of history peer out from the wall along the Hall of Honor while a face of the future explores an example of the timeless march of technology.

World War I
The Early Years

Wearing a football helmet for protection and a bicycle inner tube as a life-preserver, Eugene Ely looked none of the part of the dashing aviator when he made his milestone landing aboard USS Pennsylvania *(ACR-4) on January 18, 1911. Yet, the determined face peeking out from beneath the outlandish garb belonged to one of the finest pilots of the day who, in addition to his shipboard feats, thrilled crowds across America as a stunt flier. Fraught with danger, flying often proved a short career for those daring men who took to the skies on uncertain wings. Such was the fate of Ely, who plunged to his death during an exhibition over the Georgia State Fair Grounds in Macon in October 1911, a scant 21 months after teaching himself to fly.*

A Spirit is Born

The dark skies cast an ominous pall over the unique endeavor. Poised upon a sloping wooden deck erected aboard the battle cruiser USS *Birmingham* (CL-2) anchored off Old Point Comfort, Virginia, the biplane began its takeoff run. Leaving its perch behind, the flying machine nosed down toward the frigid waters below, kicking up spray before regaining enough speed to climb skyward. The time was 3:16 p.m. on November 14, 1910. Eugene Ely, a self-taught flier who had yet to reach his twenty-fifth birthday, had made history by becoming the first man to fly from the deck of a ship.

It would be but one step on the intrepid flier's path into aviation history. Less than two months later, on the morning of January 18, 1911, the wheels of Ely's Curtiss Pusher met the wooden deck assembled on the stern of the armored cruiser USS *Pennsylvania* (ACR-4) anchored in San Francisco Bay. Weighted by sandbags, a series of ropes strung across the landing platform brought the aircraft to a stop scarcely 60 feet from the spot where it had initially touched down, the first of many thousands of arrested landings aboard U.S. Navy ships.

Shortly after his impressive feats, Ely expressed his thoughts about the potential of aviation in the Navy. "I have proved that a machine can leave a ship and return to it, and others have proved that an aeroplane can remain in the air for a long time," he proclaimed in a letter to Captain Washington Irving Chambers, the Navy's officer in charge of aviation. " … I guess the value of the aeroplane for the Navy is unquestioned." A little less than four months later, on May 8, 1911, the sea service decided to order its first airplanes and United States Naval Aviation was born.

The Navy's first aircraft, the Curtiss A-1 Triad, took to the air for the first time on July 1, 1911. Accumulating 285 flights before being damaged beyond repair on October 16, 1912, it served as a test bed for numerous early experiments, particularly in connection with the evaluation of catapults. It was at the controls of the Triad that some of the Navy's pioneer aviators, including the service's first flier, Lieutenant Theodore "Spuds" Ellyson, honed their flying skills. The Museum's A-1 is one of two reproductions built to commemorate the Golden Anniversary of U.S. Naval Aviation in 1961.

The A-1 was characteristic of Naval Aviation's early flying machines, whose seemingly constant mechanical maladies and temperamental flying characteristics made for no such thing as a "routine" flight.
"There wasn't much sense of security for a newcomer," wrote pioneer naval aviator Vice Admiral Patrick N.L. Bellinger of flying in one of the early machines. "The fact that you sat in the clear with nothing in front of you and little around you gave the would-be passenger a lot of food for thought before making his decision to fly."

"[He] looked as though he had never been on anything faster in his whole life than a tricycle," recalled John H. Towers, Naval Aviator Number 3, of his first meeting with aircraft manufacturer Glenn Hammond Curtiss. Indeed, the quiet manner of the New York-born inventor, who gained national fame with his exploits as a motorcycle speed king before turning to aeroplane design, masked his daring soul and mechanical genius. Having built the aircraft used by Eugene Ely on his renowned flights from USS *Birmingham* and USS *Pennsylvania*, Curtiss was synonymous with the birth of Naval Aviation. In the skies over his native Hammondsport, he taught some of the first naval aviators to fly, and the little hamlet along Lake Keuka gained widespread recognition for the array of flying machines born on the drawing boards of his plant there.

Glenn Hammond Curtiss

On the heels of his success in building hydroaeroplanes such as the Triad, Glenn Curtiss took the logical step in waterborne aircraft design by building the flying boat.
Following testing of the experimental designs in 1912, he initiated development of the F-Boat, which was the Navy's flying boat trainer at the end of World War I. *Some modernized versions, like the one above, were called MF-Boats.*
"Safe as a church" was how one naval aviator described the Curtiss design after making flights in one. However, to many who flew the aircraft, the location of the engine just above their heads during flight didn't provide a feeling of security.

Patrick N.L. Bellinger was the eighth officer designated a naval aviator. A central figure in the evolution of Naval Aviation, he commanded two of the Navy's first four aircraft carriers.

Unlikely Combatant

One of the first Curtiss F-Boats procured by the Navy, called the C-3 and later designated the AB-3, made military aviation history.

In April 1914, members of the Navy's infant aviation arm joined a naval force sent to Vera Cruz, Mexico, in response to the unlawful arrest of naval personnel there. On April 25 Lieutenant Junior Grade Patrick N.L. Bellinger, at the controls of the C-3 flying boat, made two flights over Vera Cruz inspecting the harbor for mines. Despite the fact that the aircraft was unarmed, the C-3 became the first U.S. military aircraft to fly a "combat" mission.

On a subsequent flight, this time in a Curtiss pusher, Bellinger returned to base with holes in his wings, the result of enemy ground fire. The Navy airman exacted a curious revenge upon his ground tormentors during his last flight over enemy territory. Bellinger had orders preventing him from firing at the ground with a pistol, so he grabbed the nearest object and took it aloft as ammunition. The first air-to-ground ordnance used against an enemy by a naval aircraft was a bar of yellow soap!

Whether it was serving as a wartime military trainer or as the aerial mount of the barnstormers who dazzled crowds across America in the years after World War I, the "Jenny" became one of the legendary aircraft of American aviation.
Though they rolled off assembly lines by the thousands, only 200 Jennys were employed by the U.S. Navy, beginning in 1916. Despite its emphasis on seaplanes and flying boats, they remained in naval service into the 1920s.
Among those who flew the "Jenny" in naval service was a small group of U.S. airmen trained in Canada by the Royal Flying Corps during the early months of American involvement in World War I. For at least one of them, the "Jenny" provided a wild ride and left a lasting impression on his Canadian hosts.

" … On returning to my own airdrome, I misjudged the landing and succeeded in placing the machine on top of the hangar in such a manner that neither tail nor wings were touching the ground. It had much the appearance of a pen stuck in the wall. I was uninjured but much ashamed and was forced to climb down the ladder and face the officers present, among whom was Lord Wellesley, the adjutant of both camps. He greeted me with 'Oh, I say, couldn't you see the 'angar?' to which I was ready to reply but wisely refrained."

Thomas H. Chapman
Diary Entry, September 25, 1917

The entry of the United States into World War I on April 6, 1917, brought Naval Aviation to a period of unparalleled expansion. The aircraft fleet numbered just 55 assorted flying machines on the day war was declared, but swelled to 2,107 aircraft and 15 dirigibles by Armistice Day on November 11, 1918. To fill the cockpits of this force, scores of young men flocked to naval air stations along the eastern seaboard for training. They were a patriotic lot, imbued with enthusiasm, curiosity about flying and a desire to make it "over there" to fight in the skies over Europe.

These would-be naval aviators first took to the air in the cockpit of the N-9 seaplane. Though early versions were underpowered and somewhat cumbersome, the N-9 proved to be an ideal primary and advanced seaplane trainer for the Navy, which took delivery of 560 aircraft during 1917–1918. As trainers, the N-9 left a lasting impression on the flight students of the day, who often made their first solos or experienced their initial dose of "stunting" in the aircraft. During a flight with an instructor, one student climbed out onto the wing of his N-9, sat for awhile, then returned to his cockpit. "It was a foolhardy stunt, but you were expected to do it to qualify as a fearless aviator." He later wrote of the experience. "... I must confess I was scared to death and very glad to get back in the plane."

" ...Loops are the most wonderful things yet ... I did not get over the top but hung upside down for a while as I did not have enough speed. It was something to look up and see the station above me ... When the plane is upside down the engine stops with a putt putt bang bang. Then when going into the dive the motor comes on with a brrrrr and races to beat the band ..."

Alfred K. Warren
Letter to his mother
December 3, 1917

Wearing a variety of uniforms, personnel at NAS Pensacola, Florida, struggle to push an N-9 seaplane into the water for the commencement of a training flight.

The tips of its machine guns protruding atop its countenance of gleaming wood and metal, a Sopwith Camel sits at rest on the floor of the Museum.
Though scores of naval aviators received their wings during the Great War, relatively few heard shots fired in anger, and none equalled the accomplishment of Naval Aviator Number 85, David S. Ingalls. In 1918, while flying Sopwith Camels of Number 213 Squadron of the Royal Air Force, the Ohio native shot down five German adversaries in six weeks to become Naval Aviation's first ace.

" ... We saw a two-seater. I opened up and caught up with him over our lines. I got under him and looked up, but I had too much speed. I fired 3 times ... [I could see] the observer plainly shooting away but I was just out of range.
"Finally, I got in a good burst at about 25 yards. I saw a big flash, and turned and dove down ... I saw him go into a spin."

David S. Ingalls
Journal Entry
September 1918

"My job is done, and perhaps you can imagine my feelings at the result. For once in my life I have had continuous unadulterated luck and come out of it without any one of the many things that I was afraid would happen ... I wouldn't be in a hurry to do it over again."
Lieutenant Commander A.C. Read
Letter to his wife
May 31, 1919

Though born of a wartime need for a long-range aircraft to combat German U-boats, the NC flying boats were destined to make history not by dropping depth charges into the Atlantic Ocean, but by crossing its wide expanse. On May 8, 1919, three U.S. Navy "Nancy" boats, as they were popularly called, took to the skies over Rockaway Beach, New York, on the first leg of a transatlantic flight attempt. Nineteen days later, the NC-4's keel sliced into the waters of Lisbon Bay to become the first aircraft to fly the Atlantic.

The fact that the aircraft was part of an attempt to cross the Atlantic was a miracle. As she splashed through the waters off NAS Rockaway Beach, New York, and slowly rose into the air, she did so beneath wings which only seven days before had carried her into the air for the first time. Her crew worked with equipment which had only once been tested in the air. The NC-4 seemed to have beaten the odds.

However, hours into the first leg of the flight to Nova Scotia, her good fortune ceased as the buzz of first one engine, then another, was replaced by silence. After making a forced landing off Cape Cod, Massachusetts, Lieutenant Commander Albert C. Read and his crew taxied to NAS Chatham, Massachusetts. As her

two sister boats continued, the flight seemed to have left the NC-4 behind. Playing on her misfortune, newsmen dubbed her the "lame duck."

As the days passed, however, bad weather in Newfoundland grounded their comrades and kept the crew of the "lame duck" in the mission. By May 15, 1919, the day before the scheduled departure on the longest leg to the Azores Islands, the NC-4 rejoined her sisters.

The following evening the trio took off, crossing the waters of the North Atlantic by night, to land in the Azores during daylight. Only the NC-4 arrived at Horta in the Azores completing the 1,300 mile flight in 15 hours and 13 minutes. Her sister boats, forced to land in order to get their bearings, were pounded by the unforgiving sea, which sank the NC-1 and left the NC-3 in tatters.

The crew of the NC-4 completed the brief flight to Ponta Delgada, also in the Azores, where weather delayed them a week. On May 27, 1919, almost three weeks after departing the United States, she appeared in the skies over Lisbon, Portugal.

Upon her arrival, Read radioed: "We are safely on the other side of the pond ... " After almost 54 total hours in the air, the "lame duck" had conquered the Atlantic.

Though Nieuport 28s were not employed by naval aviators in World War I combat, the Navy obtained a dozen examples from surplus aircraft brought back from France by the Army. They were used to evaluate the feasibility of flying off platforms built atop the gun turrets of the service's battleships.

Though many aircraft operated from the deck of USS Langley *(CV-1), the Navy's first carrier, the TS-1 fighter was the first plane designed specifically as a carrier-based aircraft. Small in stature with a length just longer than 22 feet, the TS-1 marked the first step in the development of carrier-based aircraft.*
The example suspended in the Museum was restored from the first of two TS-2s, modified TS-1s with Aeromarine engines, built by the Naval Aircraft Factory. It was altered to the TS-1 configuration to represent that historic Navy aircraft.

When she was commissioned in 1913, few people in the Navy could have imagined that the collier USS Jupiter *(AC-3), designed to carry coal, would one day launch and recover flying machines on a flight deck erected upon her hull. Yet looking from the vantage point of history, events in her life seemed to indicate this destiny.* Jupiter's *first commanding officer was Commander Joseph Mason Reeves, who would later be designated a naval aviation observer and pioneer the development of carrier tactics during the late 1920s. Additionally, after the entry of the United States into World War I, it was aboard* Jupiter *that the first naval aviators sailed for France in 1917. Decommissioned as a collier in March 1920,* Jupiter *was recommissioned as USS* Langley *(CV-1), the U.S. Navy's first aircraft carrier on March 20, 1922. Because of her appearance, and befitting her pioneer role, the ship soon acquired the nickname "The Covered Wagon."*

"As I taxied out to take off, the first novelty was the sound of the engine ... The full power of the 400 horsepower became the sound of a high speed industrial sewing machine ..."

Brigadier General Joseph P. Adams, USMCR (Ret)

One of the hottest aircraft of its day, the F6C Hawk achieved fame in the air races that captivated the American public during the roaring twenties. Though less publicized than the other aerial speed duels of the day, the Curtiss Marine Trophy Race provided a racing forum uniquely naval in nature given its emphasis on seaplanes, and no aircraft would dominate the event as did the F6C. Hawks captured three titles during the period 1926–1930, the last being the final running of the race.

First to Fight — With the familiar Marine Corps globe and anchor adorning its side, an F7C spreads its wings over the Museum. The first leatherneck assigned to duty involving flying, First Lieutenant Alfred A. Cunningham, reported on May 22, 1912, the beginning of the Marine Corps' association with Naval Aviation.

Born in an era of wood and fabric, the Ford Tri-Motor was the largest all-metal aircraft in the United States upon its 1926 introduction. The Navy procured nine of the sturdy aircraft, and one privately owned "Tin Goose" carried a naval aviator to glory. In November 1929, Lieutenant Commander Richard E. Byrd and a crew of three navigated a trusty Ford nicknamed "Floyd Bennett" on a flight over the South Pole, becoming the first to traverse the world's southernmost point by air.

With the dirigibles serving as their "big top," the handful of pilots who flew the F9C Sparrowhawk during the early 1930s were the Navy's "daring young men on the flying trapeze." Fitted with skyhooks atop their upper wings, the F9Cs executed one of the most unique flying evolutions in the history of Naval Aviation. Beginning a flight, the aircraft were lowered from the cavernous innards of the dirigible by way of a trapeze-like contraption and released. Upon return, the pilot engaged his aircraft's skyhook onto the suspended trapeze and was pulled aboard. Though the remains of four Sparrowhawks are present amidst the wreckage of the rigid airship USS Macon *(ZRS-5) that crashed off the coast of California in 1935, this aircraft is the only one displayed anywhere in the world.*

While intended to be the "eyes of the fleet," scanning the vast expanse of ocean high above ships at sea, and the Navy's "aircraft carriers in the sky," a nest for flocks of scouting planes, the Navy's rigid airships were short-lived. Though their sheer size made them a novelty in the aviation-crazed America of the late 1920s and 1930s, their strategic value to the Navy never materialized and tragedy marred their existence. Of the five dirigibles built for the Navy, four met their ends in fatal crashes.

A Sparrowhawk pilot shows how it's done aboard the rigid dirigible USS Macon *(ZRS-5).*

Only 27 production models of the F11C-2, later redesignated BFC-2, were delivered to the Navy by Curtiss. Established as a Fleet Air Detachment in 1919, the "High Hats," now called the "Tophatters," are the oldest squadron in the history of United States Naval Aviation. They were the only outfit fully equipped with the BFC-2 Goshawk, and flew the aircraft during the period 1933–1938.

In contrast to the array of dials and switches in modern military aircraft, the fighter pilot of the late 1930s faced a less daunting instrument panel when he took to the skies.

The last of the biplane fighters produced by Grumman Aircraft Engineering Corporation, the F3F also represented the passing of an era for the United States military as the last biplane fighter procured by any service. First delivered by Grumman in March 1936, F3Fs equipped five Navy and two Marine Corps squadrons in 1940. Retired from front-line service on the eve of the United States entry into World War II, many of the remaining F3Fs served as trainers during the early years of the war.

For the aviators who flew them in training, men who would face the crucible of combat in far-flung Pacific skies, the maneuverable Grumman biplanes and their 950-horsepower engines symbolized their baptism into fighter flying. "It was just like a bumble bee, and it was one heck of a lot of fun, I tell you," one aviator fondly remembered.

The wind whipped past the cockpit and the sun gleamed off the rolling Pacific swells below. Robert Galer, a 27-year-old Marine First Lieutenant, prepared to guide his F3F-2 to a landing aboard the aircraft carrier USS Saratoga *(CV-3) operating off the California coast on August 29, 1940. Downwind of the carrier, after following the standard procedure of switching to the auxiliary fuel supply, the reassuring roar of the engine was replaced first with coughing and then eery silence.*

At an altitude of 1,000 feet, there was no alternative to ditching, and Galer set his troubled mount into the Pacific with a splash. A nearby destroyer sent a boat to pluck the soaked Marine from the sea, while the F3F-2 nosed over and descended to a watery grave. In 1990, while searching for a helicopter that had crashed at sea, a Navy submersible spotted the silhouette of an aircraft, the yellow band and numbers on its fuselage still visible.

After 50 years, Galer's biplane fighter had been found. Recovered in April 1993, the aircraft was met at a pier at NAS North Island by its last pilot, now a retired brigadier general, fighter ace and holder of the nation's highest decoration for valor, the Medal of Honor.

The only original F3F anywhere in the world, the aircraft has been a featured display at the Museum since February 1995, after restoration by the San Diego Aerospace Museum.

By 1961, none of the Grumman FF/SF series of aircraft, to say nothing of their export versions, were known to exist, and it seemed an important link in Naval Aviation history had been lost forever. However, that year Oklahoman J.R. Sirmons, a fertilizer and crop duster pilot on contract in Nicaragua, noticed a familiar aircraft silhouette in a junkyard in Managua.

Determining that it was a Grumman biplane, Sirmons paid $150 for the heap, eventually restoring it with the assistance of Grumman and flying it to the United States in 1966. Found to be a G-23, one of 52 export versions of the aircraft completed in Canada with the assistance of Grumman, the aircraft was eventually restored as an FF-1 and appropriately adorned with the markings of Fighting Squadron 5B, the Red Rippers. Eventually acquired by Grumman, the treasure was flown to NAS Pensacola in June 1967 and donated to the Museum.

First delivered to the Navy in 1933, the Grumman FF-1 marked the young Long Island company's first venture into producing naval aircraft. Featuring retractable landing gear and an enclosed cockpit, which were unique for the period, the aircraft quickly earned the nickname "Fifi," a natural result of the designation FF-1.

However, the design's squat metal fuselage and mammoth cowl ring around its engine belied this feminine moniker and exuded the durability and power that were to become characteristic of Grumman aircraft. A total of 27 "Fifis" served in front-line Navy service, most notably with the famed Red Rippers of Fighting Squadron 5B.

World War II

As much at home on water as in the air, a PBY Catalina sloshes through a calm sea.

As the 1930s drew to a close, aviation had gained a foothold in the fleet, and those who had guided it from the uncertain days in Hammondsport and Pensacola could take pride in what Naval Aviation had become. However, the light of accomplishment that had marked peacetime was eclipsed by the clouds of another global conflict.

The world war that began with the scream of German dive-bombers piercing the air over the Polish countryside changed the face of history and Naval Aviation. In the crucible of combat naval air power proved its value and effectiveness, and as the United States emerged victorious from a shattered world, Naval Aviation stood as the most powerful striking arm of the nation's sea service.

In 1940, the first full year of World War II, 708 men were designated naval aviators. By 1945, when the war ended, 49,000 Navy men and 10,000 Marines wore wings of gold. For this generation of fliers, a host of training aircraft came to symbolize the emotions experienced along the path toward wings. From the time he made his first solo flight, which one aviator described as "a feeling of satisfaction, freedom, inspiration and a little bit of awe," to that moment when he could finally call himself a naval aviator, planes with nicknames like "Yellow Peril," "Vultee Vibrator" and "J-Bird" stood at every step.

Mastery of them brought wings of gold, the tangible representation of a challenge met. In the words of a wartime letter from one cadet to his parents: "In the little piece of gold are hours of sweat, tears, thrills, disappointment, determination, happiness and many prayers. It has taught me so far like nothing else."

"From every corner and every crossroads; from the prairies, the hills; from the cities; ... from all the ends of big, wondrous America they had looked to the sky, found it a challenge, and had come."

Flight Jacket
NAS Pensacola Cadet Yearbook
1942

From the waters of Pensacola Bay to naval reserve air bases scattered across the United States to their final nesting place on the banks of the Severn River in Annapolis, N3N-3 Yellow Perils served Naval Aviation for nearly a quarter of a century. Though the end of World War II signaled retirement for most of the N3Ns, a handful of the seaplane versions of the aircraft such as this one were assigned to the United States Naval Academy to indoctrinate midshipmen in aviation. When retired from that duty in 1959, they were the last biplanes used by the U.S. military services.

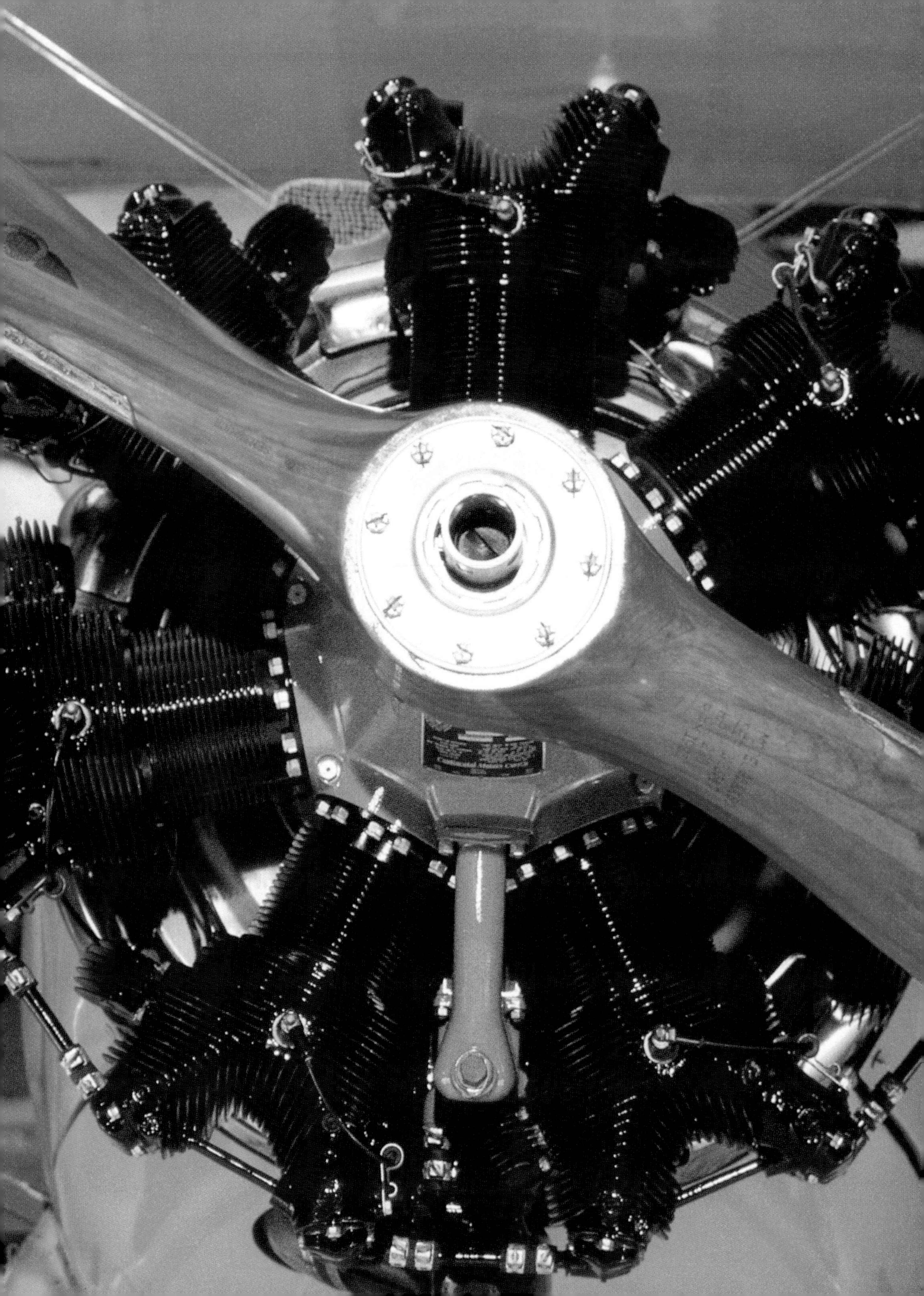

Though it was one of hundreds of Stearmans built for both the Army and Navy during World War II, what makes this particular aircraft unique is the aviator who once sat in its cockpit. On a frigid day in January 1943, an 18-year-old student naval aviator twice climbed aloft in this biplane trainer for solo flights over the frozen landscape surrounding Naval Air Station Minneapolis, Minnesota. He logged a total of 2.2 flight hours in the aircraft that day before leaving it behind and continuing along the road toward wings of gold. The aircraft eventually became a cropduster.

Acquired by the Museum in 1992, the aircraft has received a visit from its former pilot, George Herbert Walker Bush, whose path eventually led him to the White House.

Stearman N2S biplanes pictured in formation in the skies over south Texas.

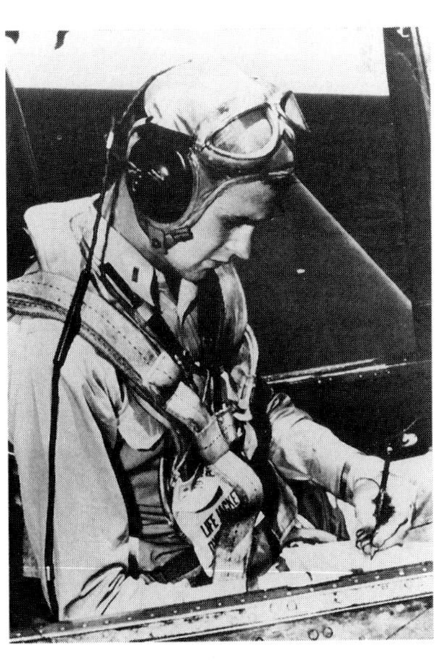

Future President of the United States George Bush pictured in the cockpit of a TBM Avenger torpedo bomber during wartime service as a naval aviator.

Facing Page: *Behind the roar of an N2S engine thousands of naval aviators first experienced the magic of flight.*

Known as one of the loudest aircraft in the sky during its years of service, the J2F Duck followed the design tradition of earlier Loening OL floatplanes by featuring a pontoon that was part of the aircraft's fuselage. The retractable wheels and a tailhook enabled the J2F to operate from carriers and shore bases as well as at sea.

Facing Page: *The yellow leading edge of its wings just visible above its gleaming underbelly, the SNC seems to glow against a background of red.*

"Those outfits [PBYs] really come down with a slam-bang when they hit the water. Sounds like you're in a bass drum … After the second hour of bounce landings a rivet was forced out of the hull and water began squirting in. They just forced a pencil in the hull and kept on flying."

Cadet Carrol Voss
Letter to his parents
April 7, 1944

The uniform lines of its wings silhouetted by a beam of light, the PBY Catalina emerges from the Museum's shadows.

With the United States' entry into World War II, the PBY quickly established itself as one of the true workhorse aircraft of the war. In addition to its value as a patrol plane, the "Cat" proved its mettle in other missions, hunting U-boats beneath the Atlantic and, disguised in black paint, launching nocturnal attacks against Japanese targets in the South Pacific. In another role, an approaching PBY was a symbol of salvation to many downed airmen floating on wide expanse of ocean.

From the pinups maintaining a vigil atop makeshift desks, to the bare board floor, the Pacific Island Exhibit re-creates the rustic lifestyle of American naval aviators who flew from the jungle airstrips of World War II.
Following page*: An island watering hole, complete with beautiful faces from home and a game of checkers waiting to be played, beckons visitors.*

Robert F. Goodspeed, a naval aviator far from home relaxes at the wooden desk upon which he pens letters to the bride he left behind.

One-Eyed Jack's
PRICE LIST
COKE - 5¢
FORGET IT!!

This colorful Grumman Wildcat is one of the few F4F-3 versions of the aircraft, which with F4F-3As and some export versions of the Wildcat, were unique for their lack of folding wings. It is displayed in the prewar Neutrality Patrol markings of Fighting Squadron 72, the squadron to which it was assigned in 1941.

LieutenantJunior Grade Edward H. O'Hare

The eight twin-engine bombers glided downward, their pilots intent on dropping bombs on the carrier USS *Lexington* (CV-2) as it sliced the waters a few miles away. Suddenly, machine gun bullets from a lone F4F-3 Wildcat fighter ripped through the engine of the rear aircraft in the formation, sending it careening toward the water trailing a plume of smoke. Turning his attention to another bomber, the Navy fighter pilot unleashed a rain of metal into the aircraft's wing, forcing it to leave the area.

Ignoring the determined fire of enemy gunners and the bursts of antiaircraft fire from the friendly ships below, the aviator methodically flamed three other enemy craft in quick succession, sending two falling into the sea and causing one to break formation. In four minutes, Lieutenant Junior Grade Edward H. "Butch" O'Hare had cut a swath through the attacking enemy, destroying three aircraft and severely damaging another pair. "[It] was just a question of gettin' in there and gettin' 'em," recalled O'Hare of the February 20, 1942 flight. "So I kept on attacking." At the time, he was credited with singlehandedly shooting down five aircraft, which made him the Navy's first fighter ace of World War II and earned him the Medal of Honor.

Though Grumman's F4F Wildcat participated in the invasion of North Africa, it achieved its greatest fame in the Pacific. It held the line during the pivotal early battles of World War II, from the epic carrier engagements at Coral Sea and Midway to the embattled skies over Guadalcanal.

The pesky Grumman more than held its own against Japan's Zero fighter, although it was inferior in speed and maneuverability.

Many of the war's finest combat pilots called its cockpit home, as evidenced by the eight Wildcat pilots who received the Medal of Honor, the nation's highest decoration for valor, for their wartime actions.

Grumman delivered its last Wildcat in 1942, but the design remained in production, designated FM, in the Eastern Aircraft Division of General Motors into 1945. With minor structural changes and a more powerful engine than its Grumman forebearers, FMs served aboard Navy escort carriers throughout the war. The Grumman/Eastern teams built a total of 7,251 Wildcats.

Fifty-nine Navy and Marine Corps pilots became aces in Wildcats during World War II by shooting down five or more enemy aircraft in aerial combat.

Aircraft from Fighting Squadron 72 carried an appropriate insignia for a squadron assigned to the aircraft carrier USS Wasp *(CV-7).*

Reminiscent of a scene aboard aircraft carriers operating in the far-flung reaches of the Pacific, the Museum's SBD-3 Dauntless receives maintenance in an exhibit re-creating the hangar deck of a World War II flat-top. A combat veteran, this particular aircraft flew missions from Henderson Field on Guadalcanal and the deck of the most decorated ship of World War II, USS Enterprise *(CV-6).*

The perforated dive flaps of the Douglas SBD, which allowed a more stable dive-bombing platform, are visible aboard this Dauntless from Bombing Squadron 5 as it sits poised to roar off the deck of the carrier USS Yorktown *(CV-10) in 1943.*

The time was less than two hours before noon in a patch of the Pacific Ocean west of Midway Island. Characteristic of the early wartime accomplishments of the nation to which it was an icon, the sun rose higher in the clear, blue sky above four mighty carriers of the Imperial Japanese Navy. Having fought off air attacks by American planes, the vessels plowed through the water, their decks full of aircraft being readied for strikes against the U.S. Navy carriers operating nearby. Many of these planes, however, had flown for the last time.

Piercing the sky above, a distinctive whining sound prefaced the blue shapes hurtling through the clouds toward the ships below. Passing through the defensive fire thrown up by Japanese gunners, the SBD dive bombers reached their drop points and unleashed their deadly cargo. In the ensuing minutes, a crescendo of explosions ripped through the air, the death knell for the carriers *Kaga*, *Akagi*, and *Soryu*. "She was a mass of flames from bow to stern, with big explosions springing up every few seconds somewhere around the deck," recalled one SBD pilot describing *Soryu*. "You couldn't see anything but the explosions."

Later that day, the American dive bombers cornered *Hiryu*, the remaining Japanese carrier, leaving her a burning wreck. Four carriers which had sailed proudly toward victory were left in ruins beneath a symbolic setting sun.

In one day, intrepid airmen flying SBD Dauntless dive bombers had stemmed the tide of Japanese conquest and for the first time resoundingly defeated the Imperial Japanese Navy in a battle at sea. It was a performance that was not atypical in the life of the venerable aircraft, which during World War II was credited with sinking more than 300,000 tons of enemy shipping, including six enemy carriers. Whether it was from the decks of flattops or from airstrips carved out of jungles from the Solomon Islands to the Philippines, SBDs served throughout World War II, earning the devotion of the pilots who flew them. "It was without doubt the finest dive bomber in the world," one of them recalled almost 50 years after last strapping into a Dauntless cockpit.

With nose pointing upward toward skies it once traversed and folded wings framing a patch of blue, the SB2C Helldiver sits at rest outside the Museum.

One of the most reviled aircraft of World War II, the SB2C Helldiver was introduced to front-line service as a replacement for the SBD Dauntless. Rushed into production and carrier operations, the SB2C was never quite able to match the success of its predecessor and instead became known more for its array of early technical maladies than its performance. Fittingly, pilots, mechanics and even the manufacturer, Curtiss Wright, called it the "Beast." Despite their less-than-glowing reputation, Helldivers operated effectively from the decks of the Navy's carriers throughout 1944 and 1945, participating in the Battle of the Philippine Sea, the Battle of Leyte Gulf and the first strikes against the Japanese home islands.

"O mother, dear mother, take
down that blue star
Replace it with one that is gold
Your son is a Helldiver driver
He'll never be 30 years old."

Stanza from World War II poem
"Helldiver Driver's Song"

The re-creation of an ordnance compartment aboard a World War II-era aircraft carrier is lined by the weaponry and tools of war.

Flight gear hangs from the bulkhead of the mock ready room as if waiting to be donned by aviators preparing to fly into harm's way.

FINDER AND IDENTIFIER
U. S. NAVY HYDROGRAPHIC OFFICE

The plexiglass chin of the K-Car reveals the controls from which World War II naval aviators maintained a vigil over convoys crossing the Atlantic, their eyes always watching for any trace of a dreaded German U-boat. Of the 241 non-rigid airships procured by the Navy between 1917 and 1958, 135 of them belonged to the K-Series. The stalwart of the Navy's lighter-than-air operations during World War II, "K-Ships" spearheaded an effort that escorted more than 80,000 ships during the period 1942-1945.

Facing page: The contents of the navigation table aboard the K-Car are arranged as if they were last touched by human hands during World War II.

The weathered appearance of the PB2Y-5R Coronado, the result of having spent many years on outdoor display, belies the spit and polish history of this particular flying boat. Outfitted as a flag transport in 1945, complete with plush quarters, the aircraft carried Rear Admiral Forrest P. Sherman, deputy to Fleet Admiral Chester W. Nimitz, to Tokyo Bay for the formal surrender ceremonies that ended World War II.

Others who flew aboard her were Admirals Thomas Kinkaid and Jesse B. Oldendorff, as well as Air Force General George E. Stratemeyer. Declared surplus following World War II, the aircraft was purchased by movie mogul Howard Hughes in 1946 for use as a trainer in preparing for the flight of his "Spruce Goose." Hughes maintained the aircraft in flying status into the 1960s. It was donated to the Museum in 1973.

The OS2U Kingfisher appears to make a banking turn to starboard as it "soars" over the Museum floor.

Upon its introduction to the fleet in 1940, the OS2U was among the last of a breed of observation aircraft whose primary mission was spotting the fall of shot from the big guns of the battle line. Though World War II offered few opportunities for the Kingfisher to practice its trade against enemy ships in a surface battle, the ungainly floatplanes served as spotters for battleship and cruiser bombardments of invasion beaches. "Old Slow and Ugly," the aircraft's unofficial nickname derived from the acronym "OS2U," also proved to be a capable air-sea rescue aircraft.

It was a Kingfisher that plucked World War I flying ace Eddie Rickenbacker and his fellow survivors from the Pacific after the plane in which he was a passenger was forced to ditch in November 1942. On April 30, 1944, an intrepid OS2U pilot named Lieutenant John Burns twice flew into Truk Lagoon to rescue carrier pilots and crewmen shot down during a raid on the island. Unable to fly the aircraft with men clinging to his wings, he instead taxiied to a waiting submarine to deposit his grateful human cargo. All told, he pulled 10 men to safety that day.

The crewman aboard an OS2U Kingfisher hooks on to the crane that will hoist his aircraft aboard the battleship USS Arizona *(BB-39). Almost two months after this photograph was taken,* Arizona *was sunk during the Japanese attack on Pearl Harbor.*

From this birdcage-like cockpit, Japanese fighter pilots created a legend. Code named Zeke by the Allies, the A6M Reisen *(Zero Fighter) first entered combat in the skies over China in 1940 and served as a front-line Imperial Japanese Navy fighter until 1945. During the early months of World War II, the A6M's superior range and unparalleled maneuverability surrounded the aircraft with a certain mystique and air of invincibility. However, as the war progressed, improved American fighter aircraft, aided by the loss of most experienced Japanese aviators in the vicious air battles at Coral Sea, Midway and over Guadalcanal, allowed the Allies to eclipse the vaunted Zero with devastating results for the Japanese Empire.*

The sun's rays shine upon the brilliant red hinomarus, *symbols of the Japanese nation the aircraft once served, adorning the wings and fuselage of the N1K2-J* Shiden Kai *(Violet Lightning Modified). One of the most capable fighters produced by Japan during World War II, the N1K2-J, code named George by the Allies, reached combat service too late to have a major impact in the air war. The aircraft served in a small number of land-based naval air groups (*kokutais*) tasked with defending the Home Islands against the final Allied aerial offensives during 1945.*

A trophy of war, this particular George was captured following the cessation of hostilities and delivered to the United States aboard the escort carrier USS Barnes *(CVE-20). After spending decades in storage at various facilities, it was acquired by the Museum in 1992. The restoration process preserved most of the aircraft's original components, including a fuselage, dented and pocked by crude repair patches and a section of the right wing punctured by an Allied bullet.*

The Goodyear version of the Corsair, the FG-1D, exhibits the unique design characteristics of the famed fighters. The Corsair was powered by a 2,000-plus horsepower engine enabling it to fly at speeds of more than 400 m.p.h. The distinctive gull wing, necessitated by the need for shorter landing gear and clearance for the massive propellers, was the root of many a nickname for the aircraft, including the "U-Bird" and the "Bent-Wing Bird."

Few aircraft in the history of aviation have matched the Corsair's longevity and effectiveness; "U-Birds" served Naval Aviation through 13 years and two wars. Squadrons with monikers such as the "Jolly Rogers" and the "Black Sheep" along with pilots named Boyington, Hanson, Walsh and Kepford made the Corsair famous during World War II. Corsair pilots were credited with shooting down 2,140 Japanese aircraft in aerial combat and, for infantrymen struggling in the coral wasteland of Peleliu and the volcanic ash of Iwo Jima, a Corsair skimming the ground on a close air support mission against enemy positions represented a true savior. This success carried over into the Korean War where the aircraft served capably in the night-fighting role and in the familiar mission of supporting ground troops.

A Corsair pilot activates the aircraft's folding wings after returning to his carrier.

The waters of Empress Augusta Bay sparkled far below as pilots of Fighting Squadron 17 maintained a watch over the ships arrayed below for the invasion of Bougainville. It would not be long before their vigilance was rewarded. Within minutes two formations of Japanese aircraft, intent on disrupting the landing craft, appeared on the scene, prompting the Navy fighters to spring into action. Emitting a distinctive whine, one the Japanese would eventually call "Whistling Death," the powerful machines doved toward their quarry. In moments, with the rattling of machine guns piercing the air, opposing pilots twisted and turned their aircraft amidst the clouds engaged in a deadly aerial dance. One by one, the Japanese fighters succumbed to their attackers, each one's demise signalled by the smoke it trailed as it plummeted toward the sea below. The seven enemy fighters shot down that day marked the beginning of a remarkable record for the pilots of Fighting Squadron 17. Nicknamed the "Jolly Rogers," they would pillage Japanese aviation in the Solomon Islands, destroying 154 aircraft in 76 days of combat. Appropriately, they would achieve this remarkable feat upon the wings of an aircraft named for the swift sailing craft manned by pirates of old, the F4U Corsair.

A motley crew of Fighting Squadron 17 personnel pose for the camera in front of the squadron tent on the island of Ondongo, November 1943. Little wonder why the squadron's unofficial nickname was "Blackburn's Irregulars" after the commanding officer, Lieutenant Commander Tom Blackburn.

A reflection of the past stretches across the floor of the Museum, its vivid colors rising to meet their source, the J4F-1 Widgeon. An unlikely combatant in the Battle of the Atlantic, this aircraft sent the German submarine U-166 to the bottom during an August 1, 1942, patrol in the Gulf of Mexico. It was the only submarine sinking recorded by a Coast Guard aircraft during World War II. The little amphibian presents a much different appearance today than it did in 1942, having exchanged its original Ranger engines for Continentals during its civilian career.

The main street of Home Front U.S.A. captures life as it was for those left behind during World War II.

Life in wartime America was like nothing before or since. While G.I.s waged war on battlefields of Europe and the Pacific, armies of civilians banded together, collecting such items as scrap iron and old rubber tires for the war effort. In factories across the nation, women took the places of men who had departed for war, their amazing productivity creating the legend of "Rosie the Riveter." Whether it was the American flag in the store window or war bond rallies, patriotic fervor abounded in a battle of good versus evil. All this combined to make the World War II home front a unique phenomena in the nation's history.

As if preparing to be lowered to the hangar deck, a TBM Avenger sits ready to fold its wings on the planks of the Museum's USS Cabot *(CVL-28) flight deck replica.*

At first it appeared as but a speck on the horizon, its presence virtually masked by a haze that hovered above the rolling swells of the East China Sea. Yet, as the lumbering aircraft droned onward, the speck was quickly transformed into a powerful behemoth, its superstructure towering high above the sea. Descending to just a few hundred feet above the sea, the warriors prepared to confront the beast, its guns winking as they fired.

Through brilliantly colored bursts of antiaircraft fire they flew, rapidly approaching the moment when they would unleash their attack. Soon bomb bay doors opened and torpedoes plunged into the water in quick succession. As the aircraft winged their way over the battleship and headed home, explosions pierced the air, signaling the end for the mighty dreadnaught. The date was April 7, 1945, and in the last great aerial torpedo attack against a ship in U.S. Navy history, TBM Avengers triggered the destruction of the mighty Japanese battleship *Yamato*, once hailed as the most powerful ship afloat.

Avengers launched lethal firepower from land and sea.

Initially produced by Grumman and later by General Motors, the TBF/TBM torpedo-bombers began their combat careers at the Battle of Midway, where five of six aircraft were lost to enemy fire. Overcoming this inauspicious beginning, the Avenger would prove an essential part of the aerial punch of the Navy's carriers, whether hunting U-boats in the Atlantic or striking Japanese targets on land and sea in the Pacific. In the latter role Avengers were flown by future President of the United States George Bush, who was shot down near the island of Chichi Jima while flying one of the TBMs.

Awash in blue and white, the F6F Hellcat rests on the mock flight deck of the light aircraft carrier USS Cabot *(CVL-28).*

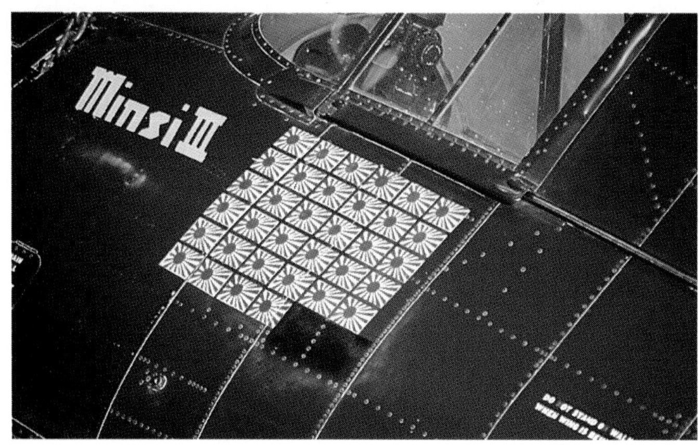

Japanese flags, symbols of the 34 aircraft shot down by Commander David McCampbell, stare out from the glossy blue skin of an F6F Hellcat painted in the markings of one of his famed personal birds, Minsi III.

On July 31, 1943, one of the U.S. Navy's foremost fighter tacticians Lieutenant Commander James H. Flatley, Jr. forecast the future of the Pacific air war, writing "Jim Flatley ... announces to Mr. Tojo that the Hellcats are taking over." One month later, as Commander Carrier Air Group Five, Flatley led a flock of F6F Hellcat fighters skyward from their roost aboard the aircraft carrier *Yorktown* (CV-10) in a raid on Japanese-held Marcus Island in the first combat mission for Grumman's new fighter. Two years later, the F6F had indeed taken over, cutting a swath of destruction that stretched to the very soil of the Japanese Home Islands.

In the hands of Navy and Marine Corps fighter pilots, Hellcats were credited with the destruction of more than 5,000 Japanese aircraft in aerial combat. In one day of combat during the Battle of the Philippine Sea, June 19, 1944, Hellcat fliers destroyed the remnants of Japan's carrier-based aircraft, downing some 300 of the enemy in an engagement known to the aerial hunters as the "Great Marianas Turkey Shoot." All told, 306 fighter aces called the cockpit of a F6F home, including Captain David McCampbell, who with 34 aerial victories stands as the Navy's top ace. So devoted to the Hellcat were its pilots that Lieutenant Junior Grade Alex Vraciu was overheard to exclaim, "If it could cook, I'd marry one."

A collage of Hellcat Aces, clockwise from upper left: Lieutenant James S. Swope, Lieutenant Junior Grade Alexander Vraciu and Commander David McCampbell.

Members of Fighting Squadron 9 pictured aboard the aircraft carrier USS Essex *(CV-9) display victories in both the European and Pacific theaters.*

"The life saving and law enforcement possibilities of the helicopter have heretofore been especially stressed. However, this machine can fulfill an even more important role, that is in providing aerial protection for convoys against submarine action, an important function of Coast Guard Aviation."

Lieutenant Commander Frank A. Erickson, USCG
June 29, 1942

Devoid of traditional wings and with ungainly rotor blades atop its fuselage, the HNS-1 Hoverfly was a novelty when introduced to the Navy in 1943. However, in the hands of the men of the U.S. Coast Guard, which during World War II operated as part of the Navy, the curious-looking machine made history as Naval Aviation's first helicopter. In addition to performing two notable wartime missions of mercy, one of which was the rescue of 11 Canadian airmen stranded amidst the snow and ice of Labrador, the HNS successfully demonstrated the feasibility of operating from ships at sea. By 1947 both the Navy and Marine Corps had established experimental squadrons for the development of the helicopter. Yet it had been the vision of pioneer Coast Guard aviators that had solidified the future of the helicopter in Naval Aviation.

Slender sweeping wings knife the air around the PB4Y-2 Privateer. With a single tail, the PB4Y-2 differed in appearance from its cousin, the twin-tailed B-24 Liberator bomber that achieved its greatest fame serving the Army Air Forces in the skies over Europe. However, the aircraft were similar when it came to performance in combat. Reminiscent of sailing privateers of old, one PB4Y-2 squadron sank 91 ships and damaged 71 others during the period April–August 1945.

Doug Kirby, of the Museum restoration staff peers into the exposed interior of the SB2U-2 Vindicator. The last of its kind known to exist anywhere in the world, the aircraft was recovered from Lake Michigan, where it ditched at the hands of a pilot practicing landings aboard the training carrier USS Wolverine *(IX-64). Introduced to the fleet in 1937, Vindicators still equipped four Navy scouting squadrons and two Marine scout-bomber squadrons at the time of the Japanese attack on Pearl Harbor.*
In a play on the aircraft's nickname, pilots called SB2Us "Wind Indicators" because of their limited performance capabilities. These would become all too apparent during the aircraft's baptism by fire at the Battle of Midway in June 1942, which would prove to be its only combat experience of the war in American service.

Dressed in the familiar colors of the "Flying Tigers" of the American Volunteer Group, complete with a menacing sharkmouth nose and Chinese wing insignia, the P-40 Tomahawk basks in the afternoon sun.

One of the most widely recognized and produced aircraft of World War II, the P-40 bore the insignia of many nations, from Europe to the sands of North Africa and the skies of Burma. In all, 13,738 P-40s — Tomahawks, Kittyhawks and Warhawks — rolled off Curtiss-Wright Corporation assembly lines between 1939 and 1944.

The P-40 gained its fame with the American Volunteer Group (AVG), a band of civilian-contract fliers hired to help defend China against the Japanese. The "Flying Tigers" flew their first combat missions on December 20, 1941. By the time they disbanded in July 1942, they had destroyed nearly 300 enemy aircraft. Among the ranks of the "Flying Tigers" were many aviators who had been recruited from the Navy and Marine Corps. Wearers of the wings of gold were among the most capable pilots in the AVG. The group's top two aces, Robert H. Neale and David Lee "Tex" Hill, and the leading aces of each of its three squadrons were naval aviators. Another AVG pilot, Gregory "Pappy" Boyington, became a high-scoring ace leading the famed "Blacksheep" of Marine Fighting Squadron 214.

The distinctive G-1 flight jackets mark the naval backgrounds of a trio of Flying Tigers as they examine a map at Kunming, China, during 1941. (Left to right) Robert B. "Buster" Keeton, Edward F. Rector and Freeman I. Ricketts were among a number of naval aviators who left the service to fight for China in the American Volunteer Group.

The windows of the cockpit akin to eyes, the face of the R5C peers into the distant skies of its past.

The bluish tint of the photograph lends an aura of mystery to a ghost from the past. Recovered from Lake Michigan where it was among the scores of naval aircraft lost in training accidents during operations aboard the training carriers USS Sable *and USS* Wolverine*, the SBD-4 Dauntless is displayed in the condition in which it spent a half century entombed in the frigid depths of the lake.*

The lines etched into his countenance marking the passage of time, a veteran enlisted Navy pilot, Stu Adams, examines a roll of honor listing his brothers in arms.

Like claws poised at the ready, motionless propellers flank the slender fuselage of the F7F Tigercat.

The period between the end of World War II and the outbreak of the Korean War brought dramatic change for Naval Aviation. Technologically, the era was one of rapid advancement marked by the introduction of jet aircraft to the fleet and efforts to provide it with a nuclear capability in the atomic age.

However, the powerful force that once marched across the Pacific began to change face, rapidly dwindling in an America transitioning to peace.

During 1944 the Navy accepted 22,004 aircraft and by the end of that year 16 fleet carriers were in commission. By 1950, only eight fleet carriers plied the world's oceans. During that year only 978 aircraft were accepted by the Navy. Additionally, with the creation of the United States Air Force in 1947, Naval Aviation became locked in a struggle to preserve a place for itself in the nuclear age. While the branches of the U.S. military sought to define their roles, trouble stirred in an Asian country far away.

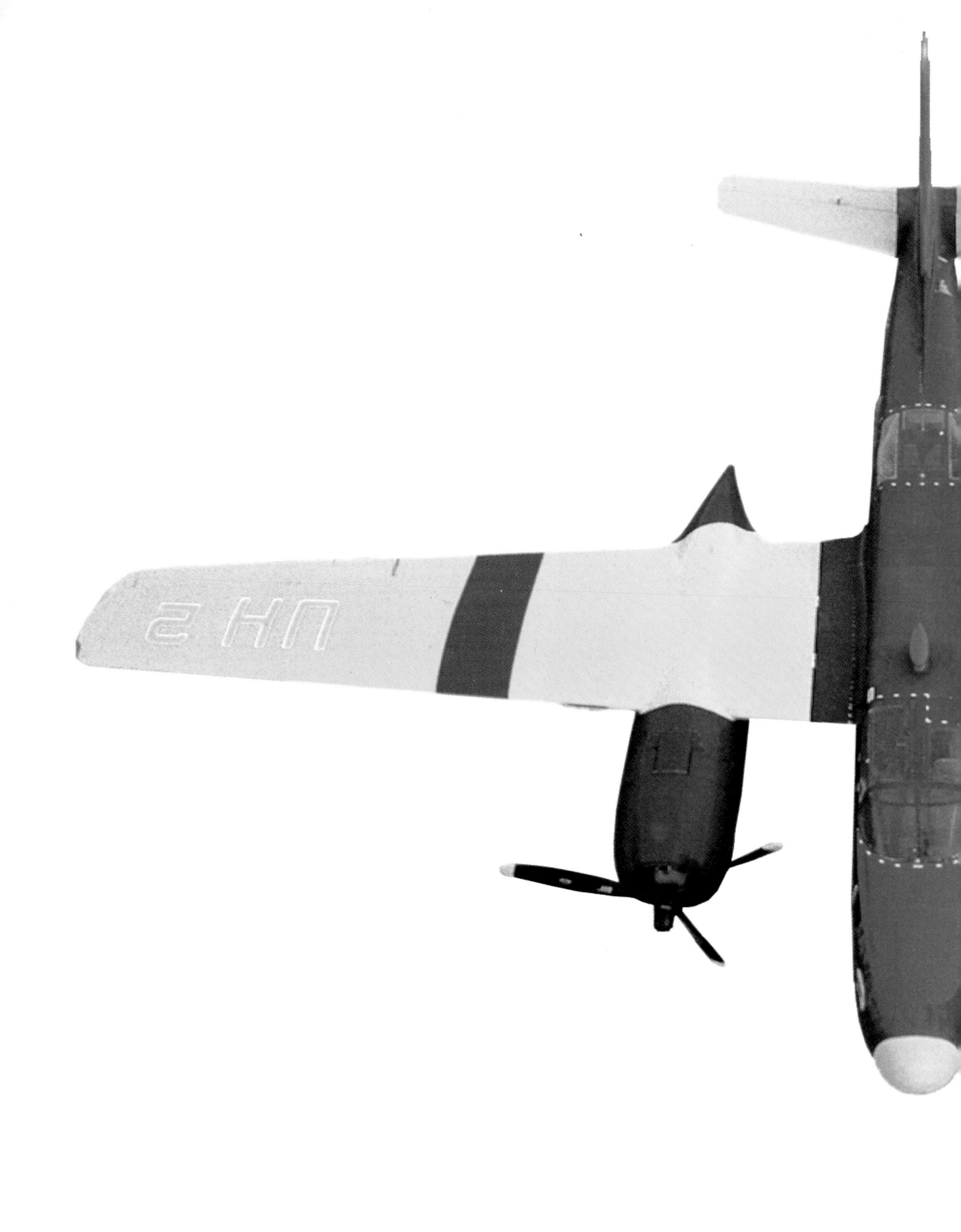
UH 2

Flanked by vividly painted wings, the mark of a drone control aircraft, the JD Invader appears ready for a mission.

"It was the closest thing to strapping on a pair of angel wings. The cockpit fit like a well-tailored glove with the canopy rails rubbing your shoulders. It had everything a fighter jock could want — speed, maneuverability, and an eye-watering rate of climb ... [It] was the last of the sports cars."
Captain Walter M. Schirra, Jr., USN (Ret)
Astronaut and naval aviator

The F8F Bearcat, was intended to blast Japanese kamikazes *from Pacific skies. The Museum's example has been beautifully restored in the markings of Commander Carrier Air Group 19. However, it arrived too late for combat with the Navy during World War II. In an era rapidly being transformed by the arrival of jet aircraft, the piston-engined F8F served for just a short period of time in front-line squadrons. The pilots fortunate enough to fly the peppy fighter never forgot the ride.*

Towering propeller blades embody the raw power of the AM-1 Mauler, which roared aloft in 1949 at a gross weight of 29,332 pounds — an amazing weight for an aircraft of its size.

The "Truculent Turtle" putters through the sky on the fuselage of its namesake.

The first production P2V-1 Neptune to roll off the Lockheed assembly line, this aircraft flew into history in the fall of 1946 in a flight from Perth, Australia, to Columbus, Ohio. Covering the distance of 11,235.6 miles in 55 hours and 17 minutes, the venerable Neptune broke the world record for distance without refueling. It took a jet-powered Air Force B-52 Stratofortress to eclipse the mark in 1962.

The crew of the "Truculent Turtle" charts the course of its record-setting flight. From left to right: Lieutenant Commander R.H. Tabeling, Commander T.D. Davies, Commander E.P. Rankin and Commander W.S. Reid.

The silver rounded shape of a jet engine's tail pipe peeks out from behind the wing of the FH-1 Phantom, Naval Aviation's first jet aircraft.

Navy and civilian personnel swarm around the D-558-1 Skystreak during an inspection of the aircraft at Muroc Army Airfield, 1947.

Set against the light of day, the shadowy silhouette of the D-558-1 Skystreak rises from a sea of darkness on a hangar tarmac.

An appropriate name for the aircraft that once held the world speed record flashes across the brilliant red fuselage of the D-558-1.

On August 20, 1947, a cylindrical-shaped aircraft, its scarlet paint scheme gleaming in the desert sun, roared off the dry lake bed at Muroc Army Airfield in California. Though dubbed the "Crimson Test Tube" by an observer, the aircraft was formally known as the Skystreak, a name which it would certainly honor on this summer day. Flashing across the a three-kilometer course with Navy Commander Turner F. Caldwell at the controls, the aircraft reached a speed of 640.663 m.p.h., capturing the world speed record for the Navy for the first time since 1923. For the Skystreak it was a shining moment in history when, upon its wings, rode the "fastest man alive."

The open-mouthed jet intake crowns the bulbous fuselage of the FJ-1 Fury against a backdrop of fiery gold. The Navy ordered 30 production versions of these stubby-nosed fighters, which lasted only 14 months in front-line squadrons.

The camera captures the long narrow wing of the HU-16 Albatross. Although it served the Navy, Air Force and Coast Guard at various times during its three decades of use, the HU-16 achieved its greatest fame in operations with the Coast Guard. Whether evacuating miners trapped in an Alaska avalanche or pulling survivors of an airline crash from the waters of the Pacific, the "Goat," as the aircraft was known to Coast Guardsmen, remained a mainstay in the air-sea rescue role even with increased use of the helicopter. Leaving that duty, HU-16s helped fight the war on drugs. It also monitored vessels at sea for illegal pollution until their retirement from the Coast Guard in 1983. This particular aircraft was the last operational Albatross.

Contrasting the classic gull-winged F4U Corsair soaring overhead, the spindly legs of the AF Guardian give the aircraft an almost insect-like appearance.

Korean War

F4U-4 Corsairs prepare to roar off the deck of the aircraft carrier USS Philippine Sea *(CV-47) for a strike on Korea, 1951. First built during World War II and in production until 1952, Corsairs flew their last combat missions for Naval Aviation in Korea.*

On June 25, 1950, North Korean forces stormed across the 38th parallel into South Korea, igniting a conventional war in which naval air was to play a decisive role. Within days the carrier USS *Valley Forge* (CV-45) joined a British flat-top in launching the first naval air strikes of the war. By the end of hostilities in July 1953, Navy and Marine Corps aircraft had logged 346,487 flights and delivered almost 200,000 tons of ordnance against enemy targets.

For three years the men and machines of Naval Aviation waged a bitter struggle in the Korean Peninsula. It was a war that was in many ways symbolic of the departure from the shadow of World War II and the entry into a new age. While carriers, aircraft and men who had served during 1941–1945 shouldered their arms once again, portents of the future were evident.

Jet aircraft and helicopters, novelties during World War II, came of age in Korea, while nations that had once been allies were now enemies. In a limited war that foreshadowed future Cold War struggles, Naval Aviation would prove its value to the nation's defense and assume a role that it would play many times in the coming years.

Smiles cross the war-weary faces of two Marine Corps fighter pilots.

Hovering just a few feet off the ground, a Marine Corps HO4S helicopter drops much needed supplies to a Republic of Korea unit cut off from friendly forces.

Ringed by the tracks of many predecessors, an F9F-2 Panther sits at rest. During 1952–1953 this particular "Cat" flew combat missions with both land-based Marine squadrons and with the Navy from the deck of the aircraft carrier USS Boxer *(CVA-21).*

A quartet of F9F-5 Panthers tilts its wings toward Mount Fuji.

The F9F Panther carried many young men into combat during its service career. Joining Navy and Marine Corps squadrons less than a year before North Korean forces marched across the 38th Parallel into South Korea, Panthers took part in the first American offensive strikes against the Communists on July 3, 1950. During these strikes, pilots of Fighter Squadron 51 downed two enemy aircraft for Naval Aviation's first kills of the war. On November 9, 1950, Commander William T. Amen, skipper of Fighter Squadron 111, became the first naval aviator to shoot down a jet, flaming an enemy MiG-15 with his Panther's guns.

F9Fs flew their share of missions, attacking land targets amidst the intense antiaircraft fire that was commonplace in Korea. Here the aircraft's ruggedness and durability brought many a pilot home safely, as described by one Marine Corps pilot known for his ability to hit baseballs out of Boston's Fenway Park.

"... I'm being called lucky by all the boys — with good cause. Some lucky bastard hit me with small arms ... — started a fire. I had no radio, fuel said empty, no hydraulic boost pressure, no airspeed .... Couldn't cut it off —[slid] on my belly 5,300 feet on fire. Why the thing didn't really blow I don't know. My wingman was screaming on the air for me to bail, but of course, with the electrical equipment out I didn't hear anything. Anyway, I was really a lucky boy ... "

Captain Ted Williams, USMCR

With floor hatch doors opened, the HUP Retriever "hovers" as if ready to pluck a downed aviator from the sea.

First delivered in 1950, the Sikorsky-built HO4S/H-19 earned its combat stripes in Korea, where it made the first large-scale combat helicopter resupply operation in history as part of Operation Windmill I on September 13, 1951. In less than three hours, the trusty helos of Marine Transport Helicopter Squadron 161 made 28 flights airlifting 18,848 pounds of cargo to an engaged Marine Corps battalion. Wearing Coast Guard colors, the HO4S was a stalwart in the search-and-rescue role until its retirement in 1966. Additionally, some versions of the aircraft were fitted with "Tug Bird" equipment, which enabled the helicopter to tow stranded naval vessels of up to 800 tons.

2
MARINES

Its name derived from the wailing roar of its engines, the F2H-2P Banshee now sits in silence. Capable of speeds approaching 600 m.p.h. and able to climb to altitudes of 45,000 feet or higher, the photo reconnaissance version of the "Banjo" proved highly successful. Triggering cameras instead of guns, pilots of one outfit, Marine Photographic Squadron 1, flew 25,000 miles a day traversing the Korean countryside on reconnaissance missions. By war's end, the squadron had shot enough film to circle the globe at the equator six and one-half times.

Though the Skyknight was intended to be Naval Aviation's first carrier-based jet night fighter, the aircraft was not used much in this role. Instead, it fell to landbased Marine Corps squadrons to demonstrate the aircraft's nocturnal prowess during the Korean War. Nicknamed the "Flying Nightmares," F3Ds of Marine Night Fighter Squadron 513 haunted Communist aircraft under the stars, shooting down six between November 3, 1952 and January 31, 1953. Major Elswin P. Dunn (left) and Master Sergeant Lawrence J. Fortin scored the squadron's fourth kill, which is recognized by the red star emblazoned on the fuselage of their F3D.

The armor of the F3D-2 Skyknight is darkened in the scheme of a Korean War night fighter.

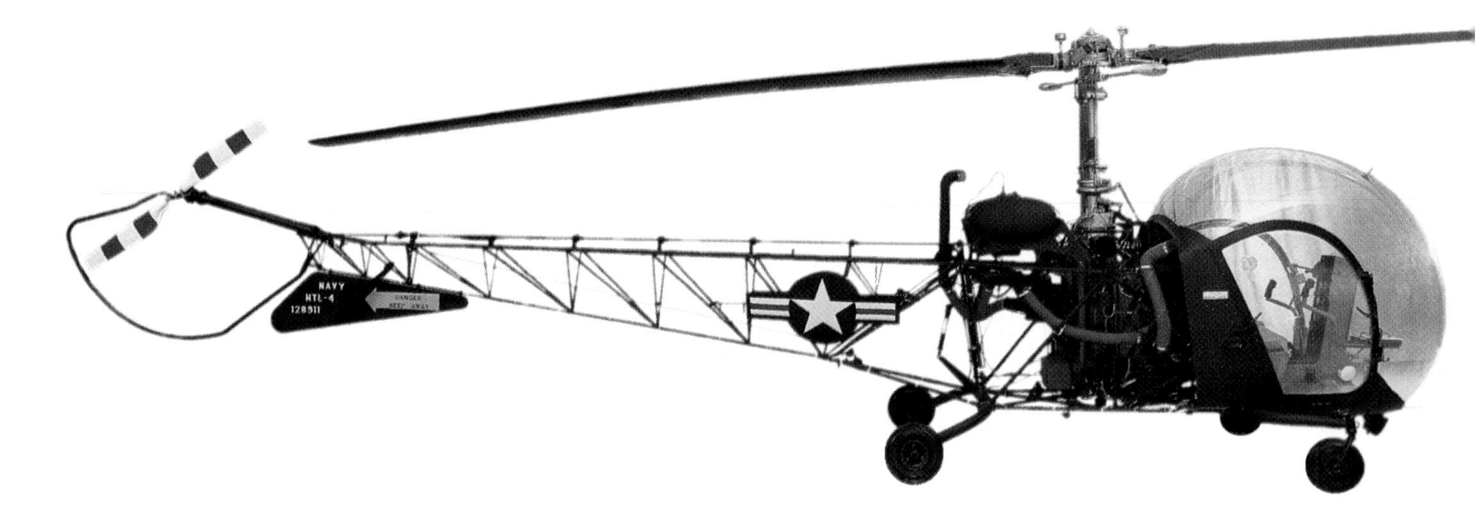

The HTL, with its fishbowl-like cockpit, was dubbed "chopper" because of the distinctive sound its rotors made when slicing through the air. It pioneered the use of rotary-wing aircraft in the medical evacuation of casualties during the Korean War. Angels of mercy to embattled infantrymen on the ground, HTL pilots saved countless lives by whisking casualties to field hospitals.
On April 23, 1951, pilots of Marine Observation Squadron 6 evacuated 50 wounded Marines, with one pilot, Captain Dwain L. Redalen, carrying 18 men to safety. All told, he spent almost 10 hours in the air shuttling casualties to the rear area.

The mighty USS Enterprise *(CVAN-65) plows through the South China Sea during one of her five Vietnam War cruises.*

On July 27, 1953, the guns fell silent on the Korean peninsula, ending three years of war. For Naval Aviation, however, the cessation of hostilities marked but the end of one Cold War battle. During the ensuing 11 years, naval air power responded to a range of crises across the globe from the waters of the Taiwan Strait to the shores of Cuba, reaffirming its place as the first line of the nation's defense.

The post-Korean era brought numerous technological advances to Naval Aviation. Jet aircraft, having come of age during the war, became larger and more capable, as did the ships from which they operated. The 1955 commissioning of USS *Forrestal* (CVA-59) opened the age of the super carrier and by 1961 the world's first nuclear-powered flat-top, USS *Enterprise* (CVAN-65), had entered service. Finally, just 50 years after the first naval aviator took to the skies, another blasted toward the heavens to become the first American in space.

A conical structure housing electronic countermeasures antennas rises from USS Enterprise's *box-like island.*

The oddly-shaped F7U-3M Cutlass grabs the spotlight. Futuristic in appearance with its tailless swept wing design crowned by a bulbous nose, the F7U also represented advancing technology in the weapons it carried. In 1956 a Cutlass-equipped squadron, Attack Squadron 83, became the first to deploy overseas with the new Sparrow I missile. However, the performance of the aircraft's engines, which one aviator remembered "generated as much heat as … toasters," plagued the Cutlass throughout its existence and gave it the uninspiring moniker "Gutless Cutlass."

The F9F-6 Cougar was the swept-wing successor of the F9F Panther.

The S-2E Tracker sits on the Tarmac splotched by afternoon rain, a folded wing revealing its inner workings.

Facing page: *As a primary, advanced and instrument trainer, the SNJ Texan was a mainstay in Navy training squadrons for nearly two decades. The Museum's aircraft is a veteran of many instructional flights from Pensacola-area fields. It is painted in the 1954 markings of Basic Training Unit Three, which at that time trained would-be naval aviators in the art that set them apart from all others — landing on an aircraft carrier.*

DON'T STALL
DON'T STALL

A stark contrast to the early birds that surround it, which never travelled faster than 150 m.p.h., the F11F-1 Tiger was the first Navy aircraft capable of flying at the speed of sound in level flight. Though Navy fighter squadrons were equipped with the Tiger for only a short time, the Blue Angels thrilled airshow crowds from 1957–1968 with the aircraft's roaring afterburner. Originally adorned in blue and gold, this particular aircraft was delivered to the Museum directly from the flight demonstration team in 1969 after having served as Blue Angel Number 1, the personal bird of the team leader.

A pair of FJ Fury fighters displays the changing colors of Naval Aviation. In 1955 naval aircraft shed their sea blue hue in exchange for a color scheme of gray and white.

Winds that once sent waves lapping against the towering hull of the SP-5B Marlin now stir only nearby blades of grass. More than 32 feet high with wings spanning 118 feet, the colossal Marlins first entered service in 1952. They were the last flying boats to serve the U.S. Navy. In July 1968 this aircraft lumbered through the air for the last time in the skies over the Naval Air Test Center at Patuxent River, Maryland, marking the final flight of a U.S. Navy flying boat.

Clouds of spray trail an SP-5B Marlin during takeoff at Guantanamo Bay, Cuba. One pilot described water operations in the big boat: "Like sitting in a chair [with] somebody hitting the bottom … with a sledgehammer with rapid, rapid blows."

International orange and white training colors adorn the barrel-like fuselage of the T-28B Trojan, a workhorse of the Naval Air Training Command for more than three decades. For prospective naval aviators moving through the training pipeline, the feeling that they had arrived swept over them the moment they sat in the cockpit and felt the rumble of power as the Trojans' 1,425 horsepower engine roared to life. For them the T-28 was their first taste of a real airplane and when taxiing down the flight line they couldn't help but look with sympathy at those in classes behind them still flying the diminutive T-34, which was known to T-28 drivers as the "teenie-weenie."

A black and yellow paint scheme likens the T-34 Mentor to a bumblebee. As the Navy's primary training aircraft since 1954, the T-34 has provided naval aviator hopefuls with their first taste of flying. The piston-engine T-34B version of the Mentor seen here has since been replaced by the turboprop T-34C.

The SNB/C-45 Navigator, a twin-engine aircraft, with a maximum speed of more than 200 m.p.h., was known universally as the "Bug Smasher" during its 25 years of service as a transport, trainer and utility aircraft.

NAVY

With exhaust pipes puffing and nose pointed down, a UH-34D leaves the deck of the amphibious assault ship USS Tripoli *(LPH-10) during a 1967 training exercise.*

Facing page*: The formally attired "Puckered Penguin," representative of Navy efforts in the exploration of the continent of Antarctica, gleams on the bright orange fuselage of the UH-34D Seahorse.*

13

Poised on the flight deck of USS Independence *(CVA-62), a "Whale" from Heavy Attack Squadron 5 prepares to launch for a combat mission off the coast of North Vietnam. Though briefly used on attack missions in Southeast Asia, the Skywarrior delivered more fuel than ordnance during the war with service as an aerial tanker. In addition, some served in night reconnaissance and intelligence roles.*

Though not the first aircraft to provide the carrier with the ability to deliver nuclear weapons, it was in the bomb bay of the A3D Skywarrior that Naval Aviation flexed its atomic muscle during the early stages of the Cold War. More than 76 feet long with a maximum weight of 82,000 pounds, the A3D was the largest and heaviest aircraft to ever operate from the deck of a U.S. flat-top, a distinction that brought with it the unofficial nickname "Whale." The jet-powered plane could carry a weapons load of up to 12,000 pounds in its belly and strike targets at ranges of almost 3,000 miles.

The towering tail of the Museum's A3D-1 Skywarrior rises toward the heavens.

Sweeping away from the fuselage, stingray-like wings dominate this view of the F4D-1. Incorporating a unique delta wing, the Skyray in 1953, became the first carrier aircraft to own the official world speed record, covering a three-mile course with a speed of 752.943 m.p.h. A quick climber, the F4D attained five time-to-altitude marks in two days of flying during May 1958. On one of these record flights, the aircraft reached 39,370 feet in less than two minutes.

Vietnam

On August 5, 1964, in response to torpedo attacks against U.S. destroyers operating in the Gulf of Tonkin, aircraft roared off the decks of the carriers USS *Ticonderoga* (CVA-14) and USS *Constellation* (CVA-64) to attack targets on the coast of North Vietnam. The action signaled the escalation of American involvement in Southeast Asia and the beginning of one of the most divisive wars in the nation's history. For almost nine years naval aviators fought in the skies over Vietnam. Whether braving surface-to-air missiles over Haiphong or enduring torture in a "Hanoi Hilton" prison cell, they served with courage and devotion to duty.

The rigid angles of the A-1H Skyraider exude a distinct ruggedness. This particular aircraft is the last Navy Skyraider to fly an attack mission, having logged its final combat flight over South Vietnam on February 20, 1968.

The "sound of power" was how one pilot described the noise generated by the Skyraider's Wright R-3350 engine.

In an effort to combine the dive-bombing and torpedo missions into a single aircraft, the A-1 Skyraider, which entered into service in 1946, was the last propeller-driven attack aircraft flown by the Navy. Capable of carrying the same ordnance load as the World War II-era B-17 Flying Fortress, a four-engine aircraft with a wingspan twice its own, the Skyraider was to Naval Aviation what a tank was to the Army.

Although A-1s flew combat missions in Vietnam, where on two occasions they outmaneuvered and shot down enemy MiG-17s, it was in America's "Forgotten War" on the Korean peninsula that the Skyraider earned its reputation as the finest attack and close-air-support aircraft of its time. Navy and Marine Corps Skyraiders flew 49,570 combat flights during more than three years of war, perhaps the most famous of which was a unique raid against the Hwachon Dam. The dam was strategically important since North Korean forces could thwart United Nations troops by flooding the Pukhan River or aid their own offensive by reducing its depth. In April 1951 senior officers ordered the structure destroyed. The task fell to AD Skyraiders aboard the carrier USS *Princeton* (CV-37). Conventional bombing attacks proved fruitless, but this did not deter the pilots of the versatile attack aircraft. Making use of a small store of World War II-era torpedoes on board *Princeton*, eight Skyraiders launched for another attack against the dam on May 1, 1951. Flying low over the water, the aircraft roared toward the structure and dropped their "fish" with a splash. Six of eight torpedoes registered hits, leaving water gushing down the face of the dam. In recognition of its success, the primary squadron that participated in the mission, Attack Squadron 195, claimed a new identity. Though called the "Tigers" since 1949, the unit quickly assumed the nickname "Dambusters."

Upon the wings of this small unassuming craft occurred one of the most dramatic stories of the Vietnam War. On April 30, 1975, during the harrowing days of the evacuation of Saigon, this South Vietnamese aircraft appeared over the aircraft carrier USS Midway *(CVA-41).*

On the third pass over the deck, the pilot, South Vietnamese Air Force Major Bung Ly, managed to drop a note indicating that he had one hour of fuel remaining and requesting permission to land aboard the carrier. With him aboard the O-1 Bird Dog were his wife and five children.

A myriad of helicopters lined the flight deck, and their crews had already departed the carrier. Midway's *Commanding Officer, Captain Lawrence Chambers, was forced to make a decision. In the time it would take to respot the helicopters, the tiny aircraft would be forced to crash into the sea. Courageously, he chose to push the expensive aircraft over the side in order to clear the deck for landing. With the carrier turned into the wind, Ly approached the stern of the ship. Though he had never seen an aircraft carrier, much less landed on one, the airman made a perfect landing to the cheers of hundreds of* Midway *crewmen.*

From Left to right: South Vietnamese Air Force Major Bung Ly brings his tiny craft to a stop aboard the aircraft carrier USS Midway *(CVA-41). Jubilant* Midway *crewmen swarm around Major Bung Ly and his wife after their flight to freedom. Ly tells his story to a journalist and* Midway *Executive Officer Commander Larry Grimes.*

FDD
5L14981

There exists no finer testament to the Skyhawk and the men who flew it than the skillful flying and courageous actions of Lieutenant Commander Michael J. Estocin. On April 20, 1967, the aviator single-handedly destroyed three enemy surface-to-air missile (SAM) sites and, despite extreme battle damage to his Skyhawk, managed to nurse his stricken craft back to a landing aboard the carrier USS *Ticonderoga* (CVA-14).

Six days later, on April 26, Estocin again confronted enemy SAMs during a strike on Haiphong, North Vietnam. Though one of the exploding missiles set his A-4 afire, he regained control of the aircraft long enough to launch anti-radiation missiles against the enemy. His crippled aircraft then plummeted into the ground. Lieutenant Commander Michael Estocin posthumously received the Medal of Honor for his actions.

Lieutenant Commander Michael J. Estocin, one of four naval aviators to receive the nation's highest decoration, the Medal of Honor, for actions during the Vietnam War.

Estocin's battle-damaged A-4E Skyhawk pictured after he landed it in flames aboard the USS Ticonderoga *(CVA-14), April 20, 1967.*

Wearing the markings of the "Saints" of Attack Squadron 163, one of Naval Aviation's most decorated squadrons of the Vietnam War, this A-4E Skyhawk seemingly roars downward to unleash its deadly cargo against an enemy target.

An A-4 Skyhawk descends for recovery aboard the floating airfield it calls home.

Nicknamed "Tinker Toy" and "Scooter" because of its diminutive stature and nimble performance, the A-4 Skyhawk was a study in simplicity in both its outward appearance and inner workings. The tiny aircraft also had an uncommon ruggedness that would come to the fore in the skies over Southeast Asia.

Participating in the first air strikes against North Vietnam on August 5, 1964, Skyhawks served throughout the war with Navy and Marine Corps squadrons, shouldering the brunt of Naval Aviation's attack effort. Dodging surface-to-air missiles over Hanoi and facing enemy small arms fire while supporting ground troops was not without costs. A total of 257 Skyhawks fell to enemy fire during the war, the most of any naval aircraft operated over Vietnam.

Almost 3,000 A-4 Skyhawks rolled off the assembly lines for America and foreign nations. Though the aircraft was retired from front-line Naval Aviation service in 1990, its familiar silhouette could still be seen in the training command in the late 1990s.

An RF-8G Crusader of Photographic Reconnaissance Squadron 206 soars high above a snowy, mountainous landscape. Photo-reconnaissance versions of the F-8 were the last Crusaders retired from U.S. Navy service.

With its wing raised and landing gear down, an F-8 Crusader resembles a hawk descending upon unsuspecting prey.

Beneath the aircraft's nose, the jet engine intake dominates the front view of the F-8A Crusader.

The F8U-1 Crusader, later called the F-8, burst onto the scene on August 21, 1956, when in the hands of Commander Robert W. "Duke" Windsor, the aircraft established a national speed record of 1,015.428 m.p.h. to capture the prized Thompson Trophy.

The next year, with future astronaut Major John H. Glenn in the cockpit, a Crusader broke the transcontinental speed mark, travelling from Los Alamitos, California, to Floyd Bennett Field, New York, in 3 hours, 22 minutes and 50.05 seconds with an average speed of 723.517 m.p.h.

Delivered to fleet squadrons beginning in 1957, the F8U was like a crusader of old, carrying the banner of the United States to Cold War battlefields worldwide. As the United States and the Soviet Union stood on the brink of nuclear war during the uncertain days of October 1962, photo reconnaissance versions of the aircraft pinpointed Soviet missile sites on the island nation of Cuba. In August 1964, F-8s joined other carrier-based aircraft in launching the first of thousands of air strikes against North Vietnam. Though the F-8's service in Southeast Asia would include photo reconnaissance and close air support, the aircraft created for itself a certain mystique in the arena of air-to-air combat. Considered by its pilots to be the last of the true fighters, Crusader drivers downed 18 North Vietnamese MiG aircraft between June 1966 and December 1968.

The Crusader survived in the naval inventory until 1987, when the last reserve photo reconnaissance squadron retired the speedy mount.

"It was the meanest looking thing on the ramp ... It just sat there [and] looked like it owned the place ... [There] was something almost evil about it."
Jack P. Brown
Former Marine Corps Fighter Pilot

F-4B Phantom IIs unleash a rain of bombs on North Vietnamese targets far below.

Its afterburners lit, an F-4N Phantom II of the Fighter Squadron 111 "Sundowners" blasts off the deck of the aircraft carrier Franklin D. Roosevelt *(CVA-42).*

One of the most successful airplanes the world has ever known, the F4H-1, which was later called the F-4, first flew on May 27, 1958, and established itself as the premier performer of its day by setting 12 world aviation records between 1959 and 1962. Though eventually employed by the U.S. Air Force, the Phantom II entered service as a naval fighter, joining its first squadron in 1960. A product of the Cold War, the F-4 was designed to serve as a high altitude interceptor, its missiles reaching out like fingers of destruction to thwart Soviet bombers attacking the fleet. However, it would be above the steamy jungles of Vietnam, that the F-4 would endure the fires of combat.

Flying from the decks of carriers in the Tonkin Gulf, the South China Sea and from air bases in South Vietnam, the F-4 was one of Naval Aviation's stalwarts of the air war in Southeast Asia. In sharp contrast to its original purpose,

From the cockpits of this Phantom, Bureau Number 153915, Lieutenant Patrick E. Arwood and Lieutenant James M. Bell of Fighter Squadron 161 shot down a North Vietnamese MiG-19 fighter during a combat mission over North Vietnam on May 18, 1972.

F-4s routinely bombed targets in North Vietnam or dropped ordnance in support of ground troops engaged in South Vietnam. Additionally, instead of long-range interception, F-4 crews found themselves engaging in dogfights with enemy MiG fighter aircraft. The Phantom scored the first MiG kill of the war for the Navy on June 17, 1965. Naval aviators ended the war with 41 confirmed victories while flying F-4s. Of these victories, five belonged to Lieutenant Randy Cunningham and Lieutenant Junior Grade William Driscoll. "The missile hit him … [and] it looked like a bunch of bees flew off," recalled Cunningham of his fifth victory on May 10, 1972. "I thought I just wounded him. I started to squeeze the trigger for my last Sidewinder [missile] and a big flame erupted … big billowing black smoke." Out of the fires of this battle, Naval Aviation produced the first American aces of the Vietnam War.

The angular tail of an RA-5C Vigilante rises to touch the rays of the sun.

A trail of fire shoots out from the exhaust nozzles of an RA-5C Vigilante, illuminating some of the darkened figures of flight deck personnel arrayed around it.

The dart-like RA-5C Vigilante, capable of flying more than twice the speed of sound, was originally designed to drop a nuclear weapon on an enemy target. Ejecting its destructive payload rearward from a bomb bay in the aft portion of the aircraft between its two powerful jet engines, the Vigilante could use its phenomenal speed to quickly depart the target area. However, it was not in this atomic mission but aerial reconnaissance that the Vigilante achieved its greatest fame. With fuselages filled with cameras, RA-5Cs streaked over the skies of Vietnam throughout the conflict in Southeast Asia, gathering vital photographic intelligence for use in prosecuting the air war.

Rotor blades droop in silent stillness, forming a crown for the H-3 Sea King. A multi-mission aircraft, the Sea King was at home lowering dipping sonar equipment into rough seas in search of Soviet submarines, or hovering over jungle clearings amidst enemy fire in efforts to pull downed aviators to safety. The H-3 has also been the backdrop for some of the most historic moments in the latter part of the twentieth century. Sea Kings pulled Mercury, Gemini and Apollo astronauts from the sea after splashdown and served as the personal helicopter of the President of the United States.

Facing Page*: Feathered propellers frame the nose of a P-3A Orion. Like the mythological creature from which it draws its name, the P-3's mission of stalking submarines hidden beneath the world's oceans continues to this day. The capability of flying on only two engines when engaged in low level antisubmarine operations allows the four-engine P-3 to remain airborne for 17.2 hours. Once a submarine is detected by the aircraft's array of sonar equipment, the Orion can launch depth charges and torpedoes against the submerged target. Though early examples of the Orion have found their way to museums, more advanced versions of the aircraft still serve as the Navy's primary land-based antisubmarine and maritime patrol aircraft.*

Captain Stephen W. Pless, USMC (far right) pictured with his crew after their daring rescue mission. (Left to right) Gunnery Sergeant Leroy Paulson, Lance Corporal John Phelps and Captain Rupert E. "Skeets" Fairfield, Jr.

It was in a UH-1E Iroquois that one of the Vietnam War's most heroic missions of mercy occurred. On August 19, 1967, Marine Corps Captain Stephen Pless, responding to a distress call, flew to a beach area where four American soldiers were under attack by an enemy force numbering more than 30 men. The U.S. G.I.s were being beaten and bayoneted by the enemy when he arrived, so Pless launched a rocket attack, killing some Viet Cong soldiers and forcing the remainder to take cover in the nearby treeline. Then, without regard to his own safety, the Marine Corps pilot landed his aircraft between the enemy force and the wounded Americans, serving as a shield for them. Taken under fire immediately, he and his crewmen beat back an enemy counterattack that reached the doorway of the helicopter. With his human cargo safely aboard, Pless skillfully flew his overloaded aircraft to safety, despite settling into the water four times as he departed the area. For his actions above and beyond the call of duty, Captain Pless received the Medal of Honor.

An HH-1K Iroquois flies across the page, its side bristling with weapons of destruction. A veritable symbol of U.S. involvement in the Vietnam War, the famous "Huey" served all United States military services in a variety of roles, including use as a troop transport, medical evacuation aircraft and gunship. In the latter role, UH-1s equipped Helicopter Attack Squadron Light Three (HAL-3) the first squadron of its kind in the history of Naval Aviation. Sweeping the brown waters of the Mekong Delta, the pilots of HAL-3 effectively supported the Navy's riverine war, destroying 1,120 structures, 1,641 sampans and 542 bunkers in one year of operations alone.

A blackened nose stands out against the predominantly international orange and white body of the TH-57C Jet Ranger, which after almost three decades still trains all Navy, Marine Corps and Coast Guard helicopter pilots.

Desert Storm
The Modern Era

A Marine Corps F/A-18 Hornet on patrol over the recently liberated sands of Kuwait shortly after the end of the Persian Gulf War.

In the early morning hours of January 17, 1991, Navy and Marine Corps aircraft took off from carrier decks and Saudi Arabian airfields to attack targets in Iraq and Kuwait. Like an orchestra playing an overture for a stage production, the crescendo of jet engines signaled the opening act of what would become Operation Desert Storm. Forty-three days later, the air campaign was over. Providing a quarter of all the aircraft taking part in the operation, Naval Aviation contributed immeasurably to the overwhelming victory in the Persian Gulf.

The success in Operation Desert Storm was a crowning moment in Naval Aviation's post-Vietnam existence, an era which like others before it brought change and accomplishment. While Naval Aviation served as the tip of the nation's foreign policy spear in places such as Libya, Lebanon, Grenada, the Persian Gulf and Bosnia, the two decades following Vietnam brought the end of the Cold War and an increasing emphasis on joint warfare.

In addition Naval Aviation continued to play a leading role in the exploration of space. When the first Space Shuttle blasted into orbit on April 12, 1981, it was manned by an all-Navy crew. Finally, the post-Vietnam era brought the designation of the first female naval aviators and an increasing role for women in the fleet.

Only the passage of time will reveal what is in store for Naval Aviation as the twenty-first century approaches. If its proud past is any indication, the future will be one marked by success in the mission of defending the nation.

With machine gun barrels protruding like fangs from the front of its slender fuselage, the AH-1J Sea Cobra appears ready to strike. A veteran of combat operations in Vietnam and Grenada, the Sea Cobra also proved a vital element of the Marine Corps air-ground team during Operation Desert Storm. Six squadrons deployed to Southwest Asia during the Gulf War. Masked in desert camouflage similar to that pictured here, Sea Cobras flew armed reconnaissance and anti-armor missions against Iraqi forces. In this capacity, Marine Corps AH-1s destroyed 97 tanks and 104 armored personnel carriers without losing a single aircraft.

Eye of the storm — A fan of bombs and missiles protrudes from beneath the wings of an Attack Squadron 46 A-7E Corsair II. A veteran of 37 combat missions during the Gulf War, this particular aircraft, Bureau Number 156804, participated in the first air strikes on Baghdad on January 17, 1991, which commenced Operation Desert Storm.

Resembling its F-8 Crusader cousin, particularly in the location of the jet intake beneath the nose, the A-7 Corsair II entered service in 1966 as the intended replacement for the A-4 Skyhawk in the light attack role. Though it didn't have the classic lines of the tiny "Scooter," the A-7 delivered outstanding performance for more than two decades. Receiving its baptism by fire in Southeast Asia, the Corsair II dropped bombs in every major crisis of the post-Vietnam War era, from Grenada, Lebanon and Libya, to the Gulf War. When retired from service following Operation Desert Storm, it left behind a legacy of accomplishment worthy of the famous name it carried, Corsair.

The reflection on a flight helmet sun visor reveals a pilot's-eye view from an A-7 Corsair II.

A diamond formation of A-7 Corsair IIs passes low over the deck of the aircraft carrier USS Dwight D. Eisenhower *(CVN-69).*

With their circular leading edges, the jet intakes of an AV-8 Harrier dominate the forward fuselage.

The camouflaged nose of the AV-8C Harrier seems to burst through the page.

The Harrier caught the attention of the United States Marine Corps in 1968. A British design with the unique ability to land and take off vertically like a helicopter, the Marine Corps viewed the Harrier as an ideal platform for close air support of ground troops because it could operate from makeshift airfields nearer the front lines. The Marine Corps ordered its first Harriers in 1969, designating them AV-8As.

It would be more than two decades before the Marine Corps' new-found aircraft entered combat for the first time. During Operation Desert Storm, AV-8Bs from five attack squadrons operated from ships and land bases. The aircraft's STOVL (Short Takeoff and Vertical Landing) capability allowed some of them to operate from a forward airfield at Tanajib, only 42 miles south of the border between Saudi Arabia and Kuwait. During the last 10 days of the Gulf War, Harriers from Tanajib flew 236 sorties, dropping 187 tons of ordnance. With rearming and refueling taking an average of 23 minutes, the aircraft could maintain a constant presence over the battlefield.

The jet blast of its engines blurs the water below an AV-8A Harrier as it descends vertically for a landing aboard the USS Franklin D. Roosevelt *(CV-42).*

Crewed by a pilot and bombardier-navigator, the A-6 Intruder provided Naval Aviation with a unique offensive punch capable of low-level bombing in any weather condition. The A-6 was introduced to combat in Vietnam just more than two years after joining the fleet, and it displayed its versatility time and again in the skies over Southeast Asia. This was demonstrated by an October 30, 1967, mission flown by Lieutenant Commander Charles B. Hunter. Hunter launched from the aircraft carrier USS *Constellation* (CVA-64) on a single-aircraft night mission against the Hanoi railroad ferry slip in North Vietnam. He evaded 10 surface-to-air missiles and antiaircraft fire to put his bombs on target. For his actions during the daring mission, he received the Navy Cross.

Following the Vietnam War, the A-6 continued to project American power throughout the world. Launching from carrier decks, the aircraft flew offensive missions against targets in Lebanon in 1983 and Libya in 1986. A total of 115 Navy and Marine Corps Intruders flew more than 4,700 combat sorties during the Gulf War. Dropping conventional bombs and firing laser-guided bombs and anti-radiation missiles, A-6s attacked Iraqi naval units and strategic targets, as well as flying close-air-support missions. Naturally, they performed day and night in any weather.

Cigar-shaped bombs hang beneath the wings of an A-6 Intruder as it glides downward toward its target during a training flight.

Against the backdrop of a setting sun, an A-6 Intruder practices touch and go landings aboard the aircraft carrier USS George Washington *(CVN-73).*

A refueling probe bisects the windscreen of the A-6E Intruder.

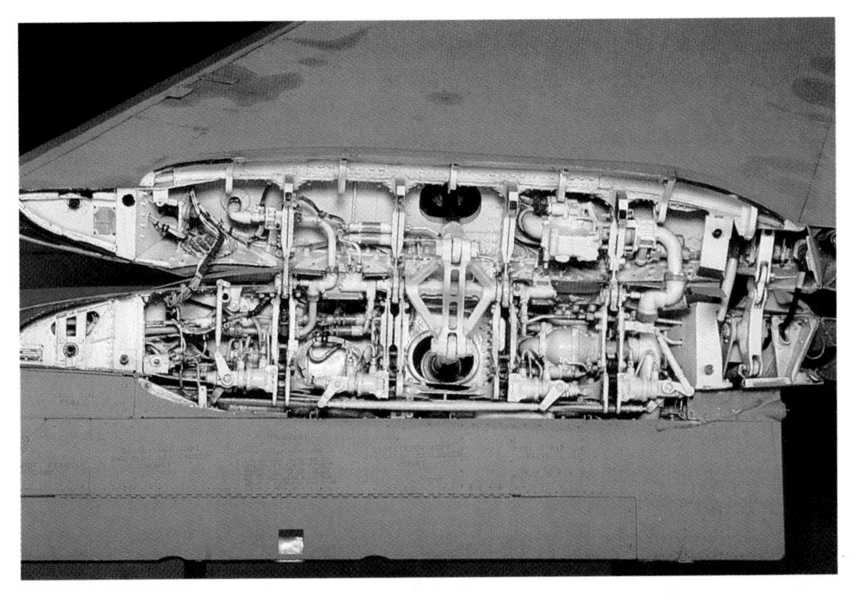

A folded wing reveals its mechanical skeleton.

Its gray fuselage scarred by operations, an F-14 Tomcat shadows the rust colored desert of Saudi Arabia.

The last of the famous "Cat" fighters produced for the Navy, the F-14A Tomcat was first delivered to Navy squadrons in 1972 to replace the F-4 Phantom II as the service's primary air superiority fighter. The Tomcat is capable of flying at speeds greater than 1,500 m.p.h. and is equipped with a variable-sweep wing that automatically shifts in flight between 20–68 degrees to provide optimum performance. The Phoenix missile and AWG-9 weapons control system carried aboard the aircraft allow the F-14 to track and attack six targets simultaneously at distances of more than 100 miles. The F-14 achieved perhaps its greatest fame on August 19, 1981. Libyan dictator Muammar Qaddafi claimed the entire Gulf of Sidra as his nation's territorial waters, a contention not recognized by the United States. Two of his air force's Su-22 Fitter fighters opened fire on a pair of F-14s from the carrier USS Nimitz *(CVN-68). In response, the Tomcats quickly engaged the Libyan aircraft, destroying both of them.*

Next page: *Reassembled just as it appeared in the Philippines, the Cubi Bar Café surrounds visitors with a rainbow of plaques, each marking the visit of a squadron to the famed Pacific locale.*

LT PHIL "PHILDO" RAIMONDO
LT JOE "FLOUNDER" THOMAS
LT TIM "SUNSHINE" DONOVAN
LTJG JOHN "TOTHEAD" TOTTEN
CWO2 RICK HEINITZ
CAPT RUSTY "SNUFFY" SMITH
WESTPAC
ON TOP, ON TIME, ORDNANCE FOR ANY OCCASION
CO LTCOL "HUNTER" BIOTY
XO MAJ "SNAKE" WHITE
SGTMAJ D.H.BARRETT
WESTPAC
JUNE-DEC 89
"ON TIME ON TARGET"
VMA - 331
HS-6 INDIANS
GROUND OFFICERS
LT DAVE ANDERSON
LT BRIAN WOOD
LT GARY HETZEL
LT SCOTT ALEXANDER
LTJG MIKE FARRELL
CVW-9
VF-24
TOP HOOKS
LT GEORGE KLESSINGER
LT.JG DAN BUNDY
CWO4 CARL COLLIER
CWO3 JIM CLARK
CRUSADERS
VMFA-122
FIRST FLEET HORNETS
WESTPAC '85
CTF
77
BLT

A panoramic view captures the pageantry of the Blue Angel Atrium, where beneath a diamond formation of blue and gold, planes of old stand aside to make way for the new.

Naval Aviation's front-line strike fighter, the F/A-18 Hornet entered service in 1980. A versatile aircraft capable of stinging both air and ground targets with a variety of weapons, the F/A-18 proved its value during Operation Desert Storm, where 138 Navy and Marine Hornets delivered more than 17,500 tons of ordnance against Iraqi targets. In a demonstration of the aircraft's multi-mission capability, on January 17, 1991, two Hornets from Fighter Attack Squadron 81 on a mission to attack an Iraqi airfield received word of enemy aircraft in the area. The F/A-18s engaged a pair of MiG-21 fighters and destroyed them with air-to-air missiles. They then delivered their bombs on target before returning to their carrier.

The same curiosity that prompted man to fly guides a tiny hand as it reaches out to touch the stinger-like nose of the F/A-18 Hornet.

Eventually day turns into night, and the long shadows of the warbirds stretch at curious angles across the Museum. The brilliant hue of a Florida sunset falls upon the statues of *The Spirit of Naval Aviation*. In one fleeting moment there is captured the exuberance of the World War II fighter pilot, his hands dancing through the air in the language of his breed, the confident stances of the Korean and Vietnam War aviators, and the pensive stares of the men of World War I and Desert Storm.

Theirs is a gathering of eagles, representative of all who pass through the ranks of Naval Aviation. It is they who have written unique chapters in the annals of flight and continue to explore the mysteries of the heavens.

Photo Credits

Special thanks to The Tailhook Association for permission to quote from The Hook magazine. Many of the historical photographs seen in this book came from the R.L. Lawson collection, which is now housed at the National Museum of Naval Aviation.

Pages 14, 16, 18, 23, 28, 33, 34, 37, 53, 61, 63, 76, 77, 89, 97, 105, 111, 113, 122, 124, U.S. Navy photos; Page 15, Curtiss Historical photo; Page 42, Photo courtesy Robert F. Goodspeed; Page 47, McDonnell Douglas photo; Page 57, Photo courtesy Tom Blackburn; Page 63, Navy Dept. in National Archives, Official U.S. Navy photo, Navy Dept. in National Archives, Navy Dept. in National Archives; Page 67, Photo donated by Robert B. Keaton; Page 76, Page 81, Major Bud Yount, USMC (Ret.); Page 86, Roland H. Baker, Jr.; Page 87, U.S. Marine Corps; Page 87, Paul Madden; Page 122, Photo courtesy Chuck Wagner; Page 124, Official U.S. Navy photo by Lt. Pete Clayton; Page 130, U.S. Marine Corps

About the Photographer Chad Slattery

"For a boy who loved airplanes, it didn't get any better than San Diego in the 1950s and 1960s."

On North Island, the Navy flung early jets off the beach. In the bay, the Sea Dart roared across the water. Convair was pushing Century series fighters out the door and designing fast jetliners. Ryan Aeronautical was over at Lindbergh Field, flying the Hummingbird and the X-13 and various exotica that ironically didn't need runways. San Diego was still small then, and safe. I biked everywhere, breathing in airplanes, mooching hops, devouring aviation books at the library.

In 1961 NAS North Island commemorated Naval Aviation's fiftieth anniversary by hosting an airshow. I begged my mom to let me use her new Kodak Brownie to photograph the planes as they flew overhead.

My folks had a glass-topped table outside. I started putting my airplane models on top of it, and photographing them looking up from below with the blue sky in back. I thought about sending the picture I made of a MiG-17 to the CIA and telling them I'd sneaked it out of Russia; apparently I had more imagination than ethics when I was 13.

In high school I decided in quick succession to become a pilot, an aeronautical preliminary design engineer, and a CIA Soviet aviation analyst. But bad eyes kept me out of the Air Force Academy, calculus traumatized me and my Russian was abominable.

So in college I studied psychology. I also edited the yearbook in my junior and senior years. The photographers had much more fun than I did. I made them show me how to use a Nikon. When I graduated I could hold my own with a pen and a camera, and I was hired by the Southern California Auto Club's Travel Publications section. On weekends I shot photos for myself. Soon I was selling stories to the *Los Angeles Times' Sunday Magazine*, to Standard Oil, and to architectural firms.

I left the Auto Club for a position as art director at a real estate investment firm in Northern California. It was my last real job. I worked there long enough to save money for what I really wanted to do: go around the world for a year. A year stretched into two. I came back with a beard and photos that started selling almost immediately.

I was 27 then. In ensuing years I worked primarily as a magazine photographer, for publications such as *Smithsonian*, *Forbes*, *Business Week*, *Time*, *Sunset*, and a host of architectural magazines.

Then in 1986 the Smithsonian Institution launched a new magazine, *Air & Space*. I sat down at the typewriter and jammed 25 story proposals onto a two-page letter; I was suddenly a 13-year-old boy again, seeing a chance to combine my favorite two things in the world — airplanes and photography. Since then I have photographed a dozen cover stories for them.

Today I photograph a wide variety of subjects, but the focus of my career is on aviation. I thrive on diversity; I have a broad curiosity and I get bored easily. Both fields — photography and Naval Aviation — are too huge to ever allow me to get bored or quit learning.